THE WELFARE OF DOGS

Animal Welfare

VOLUME 4

Series Editor

Clive Phillips, *Professor of Animal Welfare, Centre for Animal Welfare and Ethics, School of Veterinary Science, University of Queensland, Australia*

Titles published in this series:

The Welfare of Dogs

by

KEVIN STAFFORD

Institute of Veterinary Animal and Biomedical Sciences, Massey University,
Palmerston North, New Zealand

 Springer

A C.I.P. Catalogue record for this book is available from the Library of Congress.

ISBN-10 1-4020-4361-9 (HB)
ISBN-13 978-1-4020-4361-1 (HB)
ISBN 1-4020-4362-7 (ebook)
ISBN-13 978-1-4020-4362-8 (ebook)

Published by Springer,
P.O. Box 17, 3300 AA Dordrecht, The Netherlands.

www.springer.com

Printed on acid-free paper

Printed in the Netherlands.

TABLE OF CONTENTS

ANIMAL SERIES BY SPECIES: SERIES PREFACE

Animal welfare is attracting increasing interest worldwide, but particularly from those in developed countries, who now have the knowledge and resources to be able to offer the best management systems for their farm, companion, zoo and laboratory animals. The increased attention given to animal welfare derives largely from the fact that the relentless pursuit of financial reward and efficiency has led to the development of intensive animal management systems that offend the conscience of many animal users. In developing countries, human survival is still a daily uncertainty, so that provision for animal welfare has to be balanced against human welfare. Welfare is usually provided for only if it supports the output of the animal, be it food, work, clothing, sport or companionship. In reality there are resources for all if they are properly husbanded in both developing and developed countries.

This series has been designed to provide academic texts discussing the provision for the welfare of the major animal species that are managed and cared for by humans. They are not detailed blue-prints for management, rather they describe and consider the major welfare concerns of the species, often in relation to similar species or the wild progenitors of the managed animals. Welfare is considered in relation to the animal's needs, concentrating on nutrition, behaviour, reproduction and the physical and social environment. Economic effects of animal welfare provision are also considered where relevant, and key areas requiring further research.

Other titles published in the series

1. Natalie Warran, The welfare of horses
2. Eila Kaliste, The welfare of laboratory animals
3. Irene Rochlitz, The welfare of cats

PREFACE

There are about 500 million dogs in the world (Macpherson *et al.*, 2000) and only a small percentage of them live as pampered pets of the relatively wealthy, the majority live free-ranging lives in Africa, Asia and Latin America. Indeed an indicator of wealth and national development may be the number of free-ranging dogs. The life of many pet dogs is long and comfortable, but they may live in socially uninteresting environments while free-ranging dogs may live short, possibly brutal but certainly complex lives. The presence of zoonoses, especially rabies, in free-ranging dogs makes their control and perhaps ultimate extinction in many countries necessary. Indeed the control of rabies may have more of an effect on the welfare of such dogs than any desire to improve their welfare *per se*.

The physical requirements of the dog are easily met. They need a warm dry place to sleep. There is a wide variety of sustaining dog foods in many grocery stores and dogs will eat what we eat or don't eat. However, the social, exercise, and activity requirements of an individual dog is more difficult to define and to meet by busy owners. The presence of animal shelters and local government dog pounds in many towns and cities in the developed world suggests that there are many problems with the welfare of dogs in these societies.

The welfare of an animal relates to its subjective experience of life. The emotional life of a dog is probably simpler than that of a human and possibly limited to a few emotions of evolutionary consequence such as fear, anger and pleasure. The strength of these emotions may be measured using physiological and behavioural parameters that appear to be common to many mammals including humans. The biological functioning of an animal, both short and long term may also reflect its welfare (Duncan *et al.*, 1993). In the short term the effort made to maintain homeostasis and the presence of stress

ix

or distress may reflect an animal's welfare. In the longer term an animal's health, reproductive success and longevity may be influenced by its physical and mental wellbeing.

The biological functioning of dogs managed under different circumstances is difficult to assess for technical, philosophical and financial reasons and a more practical approach to discussing and assessing animal welfare is used in this book. In this approach the physical components of an animal's life such as its nutrition, health and comfort and its ability to behave 'normally' are evaluated, and its experience of suffering (pain, anxiety, fear, distress) is assessed using physiological, immunological and behavioural parameters. We can never know the subjective experience of a dog, but we can be reasonably sure of its physical condition and can use the parameters mentioned to give some indication of an animal's experience of suffering and pleasure. However, in the end this interpretation is always subjective. Choice tests and demand curve tests are used to get an idea of what is important to an animal and how hard it will work to gain access to different elements of the environment. These tests have not been used on dogs to identify their environmental needs.

The physical components of dog welfare have been intensely studied. Knowledge of canine nutrition and dietetics have increased greatly in the last 40 years and the nutrition of dogs owned by people who buy dog food is probably better now than ever (Chapter 4). The developments in veterinary medicine since 1950 mean that all the major infectious diseases and most parasite infestations of dogs can be prevented or treated (Chapter 5). Veterinary surgery now allows safe de-sexing, a major weapon in improving the welfare of dogs, and the treatment of many previously life threatening or painful conditions for companion, sport and working dog (Chapter 8). In the developed world at least there are sufficient veterinarians to tend the dog. The development of cheap and effective analgesics has been a major step towards reducing pain in dogs (Chapter 6).

The ecology of stray, free-ranging and feral dogs and their control is now well understood and if there are still many unwanted free-ranging dogs it is not due to a lack of knowledge (Chapter 2). The welfare of dogs in animal shelters (Chapter 10) and in research laboratories (Chapter 9) has been studied in some depth. The use of the dog in research has declined in some countries over the last decade, but it is a useful model for many human diseases and surgical development.

There is a paucity of research into the psychological status of companion dogs (Chapter 12). There are many problems with the management of pet dogs which will impact on their welfare. The explosion of interest in the behaviour problems of companion dogs suggests that all is not well with them (Chapter 11). The aetiology and pathophysiology of anxiety-based behaviours is poorly understood despite excellent work by dedicated therapists such as Overall (1997). The observation that pet dog numbers are declining in some countries is almost certainly good news as it may result in a decrease in the number of unwanted and unhappy dogs.

Most of the emphasis on animal welfare research has been on quantifying and reducing the deficits in the lives of animals and few attempts made to monitor the pleasure experienced by animals (Odendaal & Meintjes, 2003). The relationship between dogs and humans is complex and depends on the attitude and behaviour of individual humans and the animal's response. Understanding the welfare of dogs is always going to be a challenge, but physically dogs have never had it so good in many parts of the world and their psychological needs are being addressed more now than ever before. Dogs are used in a wide variety of activities and it is naïve to believe that their welfare will not always be compromised to some degree regardless of the intentions of owner or handler. However at present the cup of canine welfare is really half full and filling.

Kevin Stafford
New Zealand
December, 2005

ACKNOWLEDGEMENTS

I thank the University of Tokyo for allowing me time and space in which to draft this book while on sabbatical leave from Massey University. I would like to thank Professors Yuji Mori and Nobuo Sasaki, and Dr Hiroyuki Nakayama, and all the students in the veterinary ethology laboratory, for making my stay at Tokyo University so very pleasant.

At Massey University, I would like to thank Dave West, David Mellor, Peter Wilson and Vicki Erceg and a 'procrastination' of graduate students who, over the last 14 years, have helped me with ideas on animal behaviour and welfare. I thank Michelle Cook and Janice Lloyd who edited this book and helped me with my thinking and writing

This book is dedicated to Yvonne van der Veen Stafford, who has supported me throughout my career and understood my need for seclusion to write this book.

Chapter 1

THE DOMESTICATION, BEHAVIOUR AND USE OF THE DOG

Abstract: The wolf was domesticated and became the dog more than 15,000 years ago. Since then dogs have been used by humans for many purposes. Initially they probably assisted humans during scavenging and hunting and acted as sentinels. Later they were used to to guard and to drive livestock. More recently the dog is being used for an ever increasing range of activities and it has become a close companion for many people. The social behaviour of the dog makes it suitable as a companion but its need to live in a group may be compromised when individual dogs live with busy people in modern post-industrial societies. The welfare of working and sporting dogs may be compromised at work but for many of them work is a positive experience. Police, military, racing and hunting dogs may be injured during training or work. Gundogs may be shot by accident. Dogs used for fighting are often seriously injured and killed. This is an unacceptable use of dogs even if the breeds used to fight are game to do so. In some countries dogs are eaten and their welfare may be compromised by inadequate management particularly during marketing and slaughter. The welfare of guide dogs for the blind and other assistance dogs is generally high. Dogs used for showing may be modified surgically to meet breed standards. Tail docking, ear cropping and dew claw removal are common practices in some countries for specified breeds. There is a dearth of information on the longevity and health of working and sporting dogs and on the stress experienced by these animals during training and work.

1. INTRODUCTION

The wolf (*Canis lupus*) was the first animal to be domesticated, becoming the domestic dog (*Canis familiaris*). The dog has been a close companion of humans for at least 15,000 years (Savolainen *et al*., 2002) but the wolf was probably a camp follower for thousands of years before that (Clutton-Brock, 1995). The role of the dog in human development is controversial, but dogs were probably used as food, as sentinel animals and in a supportive role during scavenging and hunting. It was a support animal when other species were being domesticated and livestock farming developed. The value of the dog to humans can be seen in the rapidity with which it moved from its site of origin in East Asia (Savolainen *et al.*, 2002) to Europe and North America (Leonard *et al.*, 2002).

The dog is a member of the Canidae family of the order Carnivora. There are between 34 (Wayne, 1993) and 38 (Clutton-Brock, 1995) canid species and they have been divided using allozyme genetic differences and chromosome morphology into four groups. These are wolf-like canids including the domestic dog, South American canids, red-fox-like canids of the Old and New World and a few monotypic genera like the bat eared fox and racoon dog (Table 1) (Wayne, 1993; Clutton-Brock, 1995).

Dogs have always been working animals, but recently, especially in post-industrial western societies, their work has decreased and they have become more important as companions. However, in other societies dogs are considered unclean and are not to be touched. The contrast between the lives of a street dog in Yemen, a working sheep dog in New Zealand and a pampered pet dog in the Unites States illustrates the considerable difference in how dogs experience life. The welfare of the dog, more than that of any other species reflects its use, and the attitude and welfare of their human community (Podberscek, 1997). Dogs are not essential animals in post-industrial urban society and therefore their welfare is impacted by the wealth and philosophy of their owners. In these societies the physical welfare (health, nutrition, comfort) of dogs may be high but their lives may be dull as they are severely restricted in where they can go, what they can do and who they can meet (Kobelt *et al.*, 2003b). By contrast, the poor physical welfare of free-ranging dogs, owned or not-owned, may be leavened by their active but often short lives.

In this chapter, the domestication and the uses of the dog will be outlined as the background to modern dog-human relations. The behavioural characteristics of the dog and how they impact on canine welfare will be discussed as will the welfare of working and sport dogs.

Table 1. Categories of canid species and their geographical distribution.

Species		Distribution
Wolf-like canids (12-30kg)		
Canis familiaris	Domestic dog	Worldwide
Canis familiaris dingo	Dingo	Australia, Asia
Canis simensis	Ethopian wolf	Ethiopia
Canis lupus	Grey wolf	Holartic
Canis latrans	Coyote	North America
*Canis rufus**	Red wolf	Southern US
Cuon alpinus	Dhole	Asia
Lycaon pictus	African wild dog	Subsaharan Africa
Wolf-like canids (5-10 kg)		
Canis aureus	Golden jackal	Old World
Canis adustus	Side-striped jackal	Subsaharan Africa
Canis mesomelas	Black-backed jackal	Subsaharan Africa
South American canids		
Speothos venaticus	Bushdog	Northeast South America
Cerdocyon thous	Crab-eating fox	Northeast South America
Chrysocyon brachyurus	Maned wolf	Northeast South America
Dusicyon australis	Falkland Island Wolf	Falkland Islands
Dusicyon culpaeus	Culpeo	Patagonia
Dusicyon culpaeolus	Santa Elena zorro	Uruguay
Dusicyon gymnocercus	Azarra'a zorro	Patagonia
Dusicyon inca	Peruvian zorro	Peru
Dusicyon griseus	Grey zorro	Patagonia
Dusicyon fulvipes	Chiloe zorro	Island of Chiloe
Dusicyon sechurae	Sechuran zorro	Peru, Equador
*Dusicyon vetulus***	Hoary zorro	Brazil
Atelocynus microtis	Small-eared zorro	Brazil
Red Fox-like canids		
Vulpes velox	Kit fox	Western USA
Vulpes vulpes	Red fox	Old and New World
Vulpes chama	Cape fox	Southern Africa
Vulpes corsac	Corsac fox	Central Asia
Vulpus ferrilata	Tibetan fox	Tibet
Vulpus bengalaensis	Bengal fox	India
Vulpus cana	Blandford's fox	Southwest Asia
Vulpus pallida	Pale fox	Sahel
Vulpus rueppelli	Ruppell's fox	North Africa and Southwest Asia
Alopex lagopus	Artic fox	Holartic
Fennecus zerda	Fennec fox	Sahara
Other canids		
Otocyon megalotis	Bat-eared fox	Subsaharan Africa
Urocyon cinereoargenteus	Grey fox	North America
Urocyon littoralis	Island grey fox	Californian islands
Nycteruetes procyonoides	Racoon dog	Japan, China

Adopted from Wayne (1993) and Clutton-Brock (1995)
*Hybrid Grey Wolf and Coyote
** Also Lycalopex vetulus

2.   DOMESTICATION

The details of the process of domestication whereby the wolf came to live with humans and became the dog is unknown, but Clutton-Brock (1995) suggested that wolves may have lived in association with early hominids from the Middle Pleistocene period and that early human hunters would have tamed wolf pups, which may have bred and scavenged around human settlements. Wolf skulls with minor morphological changes suggestive of domestication have been found on archaeological sites 14,000 and 10,000 years old in Europe and Alaska respectively (Musil, 1984; Olsen, 1985). The refinement in the use of bows and arrows for hunting in the Mesolithic Period may have given the dog a role in tracking wounded animals (Clutton-Brock, 1995) and accelerated the process of domestication. The early intimacy of humans and dogs is suggested in a 12,000 year old grave in the Jordan valley in which a small dog is buried with a woman (Davis & Valla, 1978). Recently Savolainen *et al.* (2002) produced genetic evidence from the mitochondrial DNA of 654 domestic dogs that the wolf was domesticated in East Asia between 40,000 and 15,000 BP. This team of researchers suggested that 15,000 BP was the likely time of domestication and the archaeological evidence from China, Southwest Asia and Europe supports this date. This date is supported by evidence of the origin of dogs in the Americas (Leonard *et al.*, 2002), but they observed that dogs were rapidly distributed throughout Europe, Asia and America within a few thousand years of domestication and that when they arrived in America they were already genetically diverse. This had led Wayne to support 40,000 BP as the time of domestication (cited by Hecht, 2002), but alternatively, dogs could have been traded widely in the years after domestication, because they were highly valued. Behaviourally, modern *Homo sapiens* may not have emerged until about 55,000 BP (Diamond, 2002) and if the earlier date is accepted then dogs have been with us for nearly three quarters of our existence! The origin of the dog in East Asia is supported by the recent finding that the most ancient breeds (Chinese Shar-pei, Shiba Inu, Chow Chow and Akita) come from that part of the world (Parker *et al.*, 2004).

The role of the dog in human development has been exaggerated by some (Newby, 1997), but the development of improved hunting technology plus the domestication of the dog allowed for more effective hunting (Eaton, 1969). Lee (1979) found that dogs were involved in up to three-quarters of the prey animals killed by the Kung San Bushmen although dingoes were of variable help to Australian Aborigine hunters (Meggitt, 1965). The human-dog hunting team probably resulted in the depletion of large mammalian prey species. This made hunting-gathering less rewarding and encouraged the development of agriculture about 10,000 BP (Diamond, 2002). Human civilisation is based on agriculture and our use of the dog to slaughter large mammalian species was probably a significant key to our becoming farmers and subsequent urbanisation. Agricultural development and sedentary living may have accelerated the development of the dog.

The domestic dog originated from five female wolf lines but there is no evidence that the morphological variation in dog breeds resulted from wolves being domesticated in different parts of the world (Savolainen *et al.*, 2002). The process of wolf domestication remains unknown and there are two, not mutually exclusive, theories. The territorial nature of wolves may have encouraged juvenile wolves that had just left the natal pack to live on the outskirts of human encampments and to scavenge off the rubbish and faeces left by humans. Human encampments might have been unattractive to pack wolves with large territories but might have been a relatively safe haven for the young wolf. Some of these wolves would have bred successfully having not sufficiently impinged on the human existence to be killed. Wolves may have accompanied humans on scavenging and hunting trips and were found to have been of use, when competing with other carnivores at kills. Wolves would have been inadvertently selected for an ability to live close to humans.

The alternative theory is that humans tamed young wolves, found them useful, as described above, and allowed them to breed. Many hunter-gatherer peoples have pets and pet keeping might be a common behaviour of humans. However, if this was the methodology then we have to ask why other mammals were not domesticated at the same time as the wolf? The wolf has two characteristics that predispose it to domestication, namely, being a social animal and having a catholic diet. However, its territoriality may also have been an important behaviour in the process of domestication, although this is not generally considered to be a factor predisposing an animal to domestication (Price, 1984).

Domestication is a biological and a cultural process (Clutton-Brock, 1995). The biological process of wolf domestication involved wolves adapting to living in proximity to humans and the natural selection for a type of wolf that could breed when living close to humans (O'Connor, 1997). Then the cultural component engaged in selecting animals that are particularly attractive and useful. The valued wolf/dog was isolated from their wild con-specifics. The artificial selection involved in domestication was probably quite severe. When Belyaev domesticated the fox (*Vulpes vulpes*) he allowed 4 or 5% of the males and 20% of the females to breed and over 40 years produced a population in which 70-80% of the foxes were human-friendly and could be reared as affectionate pets (Trut, 1999).

The numbers of wolves killed as unsuitable during the process of domestication is unknown, but it was certainly substantial. One can imagine the brutality expressed towards a potential predator as it moved to live alongside people. Even when the usefulness of the wolf/dog had been recognised a scarcity of food might still have caused considerable malnutrition for them. In an attempt to domesticate river otters (*Lutra lutra*) and grey rats (*Rattus norvegicus*) only 16% and 14% respectively were reproductively successful and Trut (1999) concluded that domestication must put animals under extreme stress. The dog is predisposed to live alongside humans and even when opportunities allow it to go feral, as the dingo has done, it is easily recaptured and tamed. In many countries where dogs have the opportunity to go feral they are remarkably unsuccessful as hunters (Butler *et al.*, 2004) and in a North American urban environment they do not reproduce well enough to establish a wild population and disappear unless an ongoing supply of pets are abandoned on to the street (Beck, 1973). Populations of wild dogs in many countries survive by living on garbage dumps apparently bound inextricably to human resources. Thus it appears that the dogs' natural environment is now close to human society (Douglas, 2000).

3.   THE OUTCOME OF DOMESTICATION

Domestication resulted in paedomorphosis with the dog becoming physically similar to the juvenile wolf (Goodwin *et al.*, 1997). The physical changes included a decrease in body and teeth size, a shortening of the skull, a change in hair colour and tail carriage, and a decrease in the brain to body size ratio (Clutton-Brock, 1995). However, the progenitor of the dog, the Asian wolf, has a smaller brain than some other grey wolf types, so the decrease in brain size may not been substantial. A scarcity of food would limit physical size. The mandible became rounded and the eyes more forward looking. These physical changes could be a by-product of the selection for docility (Wayne, 2001) and similar physical changes were observed in the fox domestication experiments of Belyaev (Trut, 1999) and in black-footed ferrets bred in captivity (Wisely *et al.*, 2002). The dingo appears to be a reasonably defined prototype dog and many street and village dogs in Asia and Africa are middle sized, fawn or black and tan, short haired and reasonably light dogs. However in breed development some infantile physical characteristics were probably deliberately selected (Coppinger *et al.*, 1987).

Physiologically the dog reaches puberty earlier than the wolf (Price, 1984). Many female dogs can breed twice rather than once a year and dog litter sizes are greater than in the wolf (Clutton-Brock, 1995). The dingo, however, usually breeds only once each year (Thomson, 1992b). In general, male domestic dogs do not assist bitches with the rearing of pups, but all the other canids and dingo males do (Thomson, 1992b). Dogs are more promiscuous than wolves and are not monogamous.

During domestication behavioural neoteny also occurred (Frank & Frank, 1982) and the adult dog now behaves as a juvenile wolf with reduced aggression and increased submission (Price, 1999). Thus, adult dogs retain the sociability of the young wolf, they bark more than adult wolves do and they continue to play. Neoteny is associated with an increased capacity to cope with environmental change (Price, 1984). Dogs are more skilful at reading human communicative signals than wolves, a skill which puppies can achieve without much contact with humans (Hare et al., 2002). This ability appears to be the result of domestication when dogs were selected for social cognitive abilities that enabled them to understand human behaviour effectively. Thus the wolf/dogs that could read human intentions were more likely to survive and reproduce successfully than their less skilful cospecifics (Mikosi et al., 2003). Wolves do not have this skill and neither do New Guinea singing dogs (Koler-Matznick et al., 2003), which appear to have lost the skill when they went feral.

Dogs exhibit a range of attachment behaviours towards their owners similar to those seen in infant humans and chimpanzees towards their mothers (Topal et al., 1998; Prato-Previde et al., 2003) suggesting a high level of attachment. This level of attachment to humans suggests that dogs regard humans as conspecifics and is the outcome of humans favouring dogs that were social and attached to them. This closeness has led to dogs being able to change from being leader to follower and vice versa, depending on the role they have with humans, as seen in guide dogs for the blind (Nadri et al., 2001) and hunting dogs. Dogs that were stroked, scratched, talked to and gently played with by their owners had increased plasma dopamine levels, which indicates that they were experiencing pleasure in this intimate contact (Odendaal & Meintjes, 2003).

The dog has been a remarkably successful species. There are about 500 million dogs (Matter & Daniels, 2000) and they have inhabited all the continents, even Antartica where they were used as sled dogs. This success is in marked contrast to their Canidae relatives, many of which including the Ethiopian Wolf (Canis simensis) are endangered. Moreover, even the progenitor of the dog, the wolf, is extinct in much of its former range and endangered in much of the rest. The ability of the dog to be useful to humans has made it one of the mammalian success stories and even when its welfare is compromised it still exists in large numbers. Local populations of dogs depend on factors such as resource availability, climate, geography but most importantly, the human attitude towards dogs (Matter & Daniels, 2000).

4.   THE BEHAVIOUR OF DOGS

The behaviour of dogs is basically similar to that of their progenitor the wolf but it has been changed during domestication. The dog is defined behaviourally as a cursorial hunter, promiscuous, a scavenger, a social animal, territorial with a propensity for barking and one that uses urine and faeces in communication and uses its teeth in aggression. The social, reproductive and foraging behaviour of free-ranging, feral and wild dogs has been reviewed recently by Matter & Daniels (2000) who identified six classes of dogs based on the closeness of the relationships with humans (Table 2).

Table 2. Classification of dogs according to their origin, behaviour and relationship with people.

Wild	Wild for thousands of years (e.g. dingo)
Feral	wild for a few generations
Free-ranging (un-owned)	not owned (abandoned or born from free-ranging female)
Free-ranging village (neighbourhood owned)	owned by villager rather than individual household and not restrained
Free-ranging family hold owned)	owned by individual household but not restrained
Restrained	owned and with restricted movement

Adapted from Matter & Daniels, 2000

The social behaviour of these different classes of dogs is influenced by the level of restriction imposed on them by their human owners, by the resources available to them (food, shelter) and by pressure of predation by humans and other species. Thus, restrained dogs live in the social group owned and determined by their owner.

Free-ranging but owned dogs often live solitary lives and rarely group up with dogs other than those with which they are familiar from their home. In several studies of North American urban and rural owned free-ranging dogs the tendency was to live solitarily and to avoid others (Beck, 1973; Daniels, 1983b). When groups formed they were transitory collections of owned and un-owned dogs (Rubin & Beck, 1982) which formed and split apart regularly (Daniels, 1983a). Group formation of owned but free-ranging dogs is strongly influenced by the behaviour and practices of the dogs' owners (Daniels & Beckoff, 1989a). As most free-ranging dogs, owned or not, live on garbage there may be few advantages to be gained from foraging in a pack. However, village dogs (free-ranging and un-owned) in one Italian study (Macdonald & Carr, 1995) lived most of the time alone but belonged to social groups as indicated by amicable interaction. These groups appeared to defend territory and there were aggressive interactions between different groups. Groups of free-ranging dogs occur in many countries.

Feral dogs tend to have larger social groups than owned free-ranging dogs. On the Galapagos Islands feral packs of up to 8 animals were found by Kruuk & Snell (1981) and in Alabama, USA, packs of from two to six animals were identified by Causey & Cude (1980). In Italy, feral dogs were generally seen in groups (Macdonald & Carr, 1995) and these groups were aggressive towards each other. Pack hunting by feral dogs is generally unsuccessful and thus they tend to live off garbage which does not require pack cooperation. Feral dogs may sometimes attack people usually children or old people and in some circumstances, pack living may protect individual dogs from larger predators and perhaps humans. The pack structure of feral dogs may lack the social cohesion seen in dingo packs, but this may improve over generations if the pack is successful in rearing puppies. The pack behaviour of feral dogs suggests that given time, and without human control, dogs will return to a pack social life. In one Italian study, a pack of feral dogs was maintained by immigrants and remained physically close together even when the den site of a female was far from the usual resting area (Boitani et al., 1995). The poor reproductive success of feral dogs is explained by the poor integrity of the pack when it comes to supporting a bitch rearing puppies. Reproductive success may be better in dingo packs that hunt, rear young and protect territory together.

Unlike free-ranging and feral dogs, dingoes live in packs that exist long-term although all members may not be together all the time (Thomson, 1992c). The pack consists of a pair and their offspring from different years. Pack size varies and many dingoes remain solitary either permanently or split from their pack temporarily. Dingoes are successful because they cooperate while rearing young (Corbett & Newsome, 1975). The social organisation of dingoes varies depending on the prey species, climate, habitat and interaction with humans (Corbett, 1995). Dingoes hunt together when targeting larger prey species such as kangaroos, but the maximum pack size appears to be six (Thomson, 1992d). Hunting packs are more successful than individuals, but when large prey become scarce then packs will split up and individual animals will target smaller species. Dingoes occur in Australia, and in many East Asian countries such as Thailand, Malaysia, the Phillipines, Laos, Indonesia and India, but, the pure dingo is in decline in all these countries as they interbreed with free-ranging dogs (Corbett, 1995).

Thus the further dogs are removed from direct human control the more likely they are to form groups and these may eventually evolve into cohesive packs. Individual free-ranging animals once released from human control can migrate into feral groups (Boitani et al., 1995) which may become self-sustaining depending on available resources and the impact of human control efforts.

Dingo puppies can be captured and tamed easily as can un-owned free-ranging dogs. However, there is little to indicate a voluntary movement of dogs born from free-ranging females back towards close contact with humans. This may be due to the danger involved, as many such animals are killed, but may also reflect the suitability of the free-ranging life living off human refuse for the domestic dog. The tendency of free-ranging un-owned dogs to form groups reflects the innate social behaviour of these animals, which has implications for the welfare of the isolated indoor dog that lives with a busy person or any dog that spends a large proportion of its time alone.

Many of their behavioural characteristics have made the dog an ideal assistant in human endeavours. The territoriality and tendency to bark have allowed us to use dogs as sentinels and guardians of property and livestock. Their scavenging has allowed us to use them for garbage disposal. Their hunting has been adjusted to suit a range of hunting conditions and live-stock driving. Their chronic sociability has allowed dogs to be exchanged regardless of age and to be introduced to new environments without diff-culty. This sociability allows dogs to fit easily into human families. The playfulness of dogs makes it easier for them to interact with humans who are also playful. In addition their catholic diet and promiscuity have allowed them to survive in urban and rural environments even when they are not owned and sustained directly by humans.

However, some of these behaviours have made the dog's existence on the street unacceptable for social and welfare reasons (Serpell, 1995). Its promiscuity and the production of surplus and unwanted pups has made reproductive control necessary for many modern urban human communities. Barking is unacceptable to urban and suburban communities as is the random depositon of faeces on streets and in parks. The tendency of some dogs to hunt other companion animals and livestock is unacceptable as is its disposition to chase after bicycles, joggers and cars.

In the English language when 'Dog' is added to any word it signifies worthlessness; as in dog rose. This indicates that humans are ambivalent about dogs, thinking of them as mean and low whilst also praising their values at hunting and other activities. Bringing dogs indoors has certainly reduced the social problems caused by dogs in public places but this practice may have resulted in problems most appreciated by dog owners and their neighbours. Its intense social nature makes living in isolation difficult and being alone for long periods of time may predispose a dog to anxiety-based behavioural problems.

In post-industrial wealthy urban societies legislation regarding dog control and ownership is making the ownership of dogs more difficult. In such societies, dogs cannot roam the streets, owners must clean up after their dogs in public places, dogs must be exercised on a leash or in specific exercise areas, dogs of specific breeds must be muzzled in public and dog breeding may be controlled. Thus many of the characteristics which made the dog such a useful companion in the past have made it a pariah in a modern urban context. The legislation restricting dog ownership and management has occurred largely because of the constant public pressure that dogs should be controlled and owners should be responsible for that control, and that dog ownership is earned and not a right *per se*. This emphasis combined with the growing concern about dog attacks has forced legislators to restrict the lives of dogs and their owners.

Legislation and social pressure to move dogs indoors to live alone for most of the time in socially uninteresting and physically dull environments has made the disparity between the nature of the dog and its everyday environment more obvious. That the dog is inherently unsuited to this isolated existence because of its behavioural requirements is probably being recognised by society and may have encouraged the decline or stabilization of dog populations in some European countries including Germany, the Netherlands and the United Kingdom. In the United Kingdom the dog population has declined from about 7.2 million in 1987 to 6.1 million in 2002 (Anonymous, 2003e). This decrease is to be welcomed as it may signify that people have recognised the unsuitability of dogs to the modern urban environment and to modern lifestyles.

In many countries, however, dog numbers are stable or increasing. The apparent increase in the dog population in China, Indonesia, or Malaysia may represent a change in the dog-human relationship as these countries become wealthy and may represent an increase in ownership rather than dog numbers *per se*. This may be accompanied by a decrease in un-owned dogs. Databases of dog populations may be quite inaccurate depending on the methodology used to estimate the population (Patronek & Rowan, 1995). It is difficult to obtain accurate data on the population of dogs in most developing countries. The World Health Organisation (WHO) databases provides figures for the numbers of dogs vaccinated against rabies worldwide and estimates of the national dog populations but these figures appear to be rough estimates.

Where there is human conflict and the rule of law breaks down or where serious human poverty exists then many owned dogs live free-ranging lives on the street and there may be many ownerless and feral animals also. Pictures of scavenging, often diseased, dogs upset many people living in countries where dogs have been moved off the streets but these dogs are living as dogs have lived for thousands of years and by definition, therefore, more normal lives, than those confined in an apartment.

5.   THE WELFARE OF WORKING AND SPORT DOGS

The majority of dogs living in Europe, North America and Oceania are pet animals although unusually in New Zealand, with its large sheep to human ratio, working dogs outnumber pet dogs 2:1. During and following domestication and human settlement dogs were used for a limited number of purposes. However, their uses increased in the subsequent millennia and continue to increase so that today the domestic dog is used for a wide range of work and sporting activities. The traditional jobs like guarding, hunting, sledding and herding continue, but nowadays dogs work with the disabled, in conservation (Engeman et al., 2002), with the military, police and customs, in search and rescue, even controlling wildlife at airports (Carter, 2003) and in the media. Dogs are involved in a great number of sports including racing, hunting of many sorts, sheepdog gundog and guard dog trials, sled racing, obedience tests of all sorts, agility, showing and newer dog sports like frisbie. Recently interest has developed in using dogs as diagnostic tools in human and veterinary medicine.

The welfare of fighting and baiting dogs is obviously seriously compromised. In addition, the welfare of other working dogs may be compromised in a number of ways which relate to their specific activity. Some jobs such as landmine detection are obviously dangerous but police dogs may become stressed and jobs like sledding are obviously hard work. Working and sporting dogs are usually bred for their particular activity and are predisposed to engage in that particular activity.

The welfare of dogs used for the different purposes varies considerably and it is difficult to discuss the degree, if any, of welfare compromise due to a lack of information and detailed research into the activity. Many dogs work under difficult physical conditions and their working life may be short which might indicate poor welfare. There is limited published information on the percentage of dogs entering training that succeed as working or sport dogs, their subsequent health and longevity and the stress caused by training and the work or sport per se. There is some data on the incidence of work-related injuries and diseases. However a brief description of the activity and the obvious dangers gives some indication of the welfare compromise, either physical or psychological. Many dogs have been bred for specific functions and engaging in these is a powerful positive reinforcement and therefore, by definition, pleasant. Also many dogs are trained to engage in specific activities by positive reinforcement and the activity is often, apparently, a pleasure. Physical dangers and injury have to be valued against the obvious enjoyment some dogs get in doing specific work.

The stress or distress experienced during training for, or engagement in, a particular activity has been poorly determined in the dog. There is a need for more knowledge about the physiological status of dogs during training and work. There are techniques available to monitor blood pressure (Vincent et al., 1993) and heart rate remotely (Vincent & Leahy, 1997). Saliva (Kobelt et al., 2003a), plasma, and faecal corticosteroid and plasma catecholamines levels are easily measured (Beerda et al., 1996). However, it may be difficult to differentiate, using physiological parameters, between the stress of the excitement of work, as seen in working sheep dogs and the distress caused by overwork. There is a need for longer term studies to examine chronic stress in working dogs by monitoring the effect of training and work on immunological status (Beerda et al., 1997) and health (Davis et al., 2003b).

Overt behaviours have been identified which occur during stress. These include paw lifting, yawning, snout licking, lowering body profile, and vocalising while more severe stress may induce dogs to pant and increase salivation (Beerda et al., 1997). Other behaviours such as stereotypic activities may indicate chronic stress but so do increased paw lifting, yawning, increased activity, nosing and urinating (Beerda et al., 2000).

In the following section the welfare issues identified with the different uses of dogs will be considered.

5.1   Guard dogs

Dogs have always been used as sentinel and guard animals and many are used to protect property and individuals from humans or other animals, and as such their work may be dangerous. Dogs have three sleep-wake cycles per hour during the night and these make them useful as guard dogs (Adams & Johnson, 1994a, b). Guard dogs may patrol inside fences or buildings or may be tied up to act as sentinels. Guard dogs may be poisoned and many are trained not to eat except what the handler gives them. Guard dogs are trained using an attack sleeve to bite and hold. Side to side movement of the sleeve by the handler places undue stress on the teeth which may fracture (Jennings & Freeman, 1998). Some breeds such as the Belgian Malinois may be prone to tooth damage. Root canal treatment which leaves the dog with a broken tooth, and placing metal crowns on broken teeth are effective therapies for broken teeth (van Foreest & Roeters, 1997). Damage caused by choke chain misuse is sometimes seen in guard dogs. The damage may involve loss of hair, skin abrasions and serious musculoskeletal damage to the neck (Jennings & Freeman, 1998).

Patrol dogs are often released to guard a property alone at night. They may have little interaction with any other dog and minimal time with the handler. Many are tied up by a chain on the property during the day and may become agitated and injure themselves. However, there is little evidence to show that penning a dog is better than having it tethered (Yeon *et al.*, 2001). There are concerns about the techniques used to train guard dogs (Chapter 7).

5.2   Dogs as food

The eating of dogs is common in China (Cui & Wang, 2001), North and South Korea and the Philippines and less frequently in Vietnam, India, Thailand, parts of West Africa (Eze &d Eze, 2002) and elsewhere. In the past the eating of dogs was widespread in Central America (Tykot *et al.*, 1996) and Europe. Some dog breeds such as the Chow Chow may have been developed as meat dogs and apparently St. Bernard Dogs are being cross bred with local Chinese dog breeds to produce a fast-growing meat dog. It is believed that millions of dogs are eaten annually but data are scarce and in Korea and parts of China dog meat is considered a health food. In South Korea 28% of people eat dog. Taiwan and the Philippines have banned the eating of dog (World Society for the Protection of Animals, 2004). Dog meat was eaten by the poor in Cambodia but recently it has become more popular and apparently paws and kidneys are particularly sought after (Prasso, 1993).

The management of dogs bred and reared for meat and their transport and management pre-slaughter and subsequent slaughter may have serious implications for their welfare. In China there are dog farms which specialise in producing dogs for the meat market. Dogs, along with many other animals and birds, are often transported to market in small cages and held in them while awaiting purchase. Slaughter may be by clubbing followed by ensanguination. This is probably no different from how many other species of animals and poultry are managed in these countries. If dogs reared in groups for meat production are healthy, fed well and given sufficient space, exercise and attention and provided they are transported and slaughtered in a humane manner then their welfare requirements are taken care of and may be no more compromised that any other meat animal.

Among the Sioux Indians of North America and in pre-colonial Hawaii, dogs that were for eating were treated well and killed quickly and with respect (Serpell, 1995). Australian Aborigines ate dog only when other food was scarce because they considered it unpalatable (Meggitt, 1965). Those who keep dogs as companions may think that the eating of dog meat is an unethical and immoral activity. However, although the use of dogs as meat is repugnant to many people this use *per se* has little welfare significance if the animals are managed correctly throughout the process.

5.3   Hunting dogs

There are many different kinds of hunting dogs; hounds that hunt in packs or pairs, gazehounds that course, terriers and daschunds that hunt above or below ground, bird dogs that point set retrieve or drive, and dogs that bail up or kill animals. Hunting dogs are an important tool in conservation being used for the detection of endangered birds, such as Kiwi in New Zealand, and for the selective detection and often destruction of unwanted and destructive pest species such as goats on the Galapagos Islands. Most types of hunting may be physically taxing but not dangerous. Hunting dogs are usually bred, trained and exercised to engage in their specific hunting activity. Injuries sustained during hunting may occur because of the enthusiasm of individual dogs.

Some specific injuries are more likely to occur in hunting dogs than others. Injuries caused by seeds, such as wild barley seeds, which may lodge in the nose and cause epistaxis, in the ears and cause otitis externa, or under the axilla or between the toes and cause ulceration, are not uncommon in hunting dogs. The aspiration of plant material was implicated in chronic suppurative and pyogranulomatous lesions in hunting dogs (Frendin, 1998). Injuries may be caused by barbed wire or thorns. Hunting dogs may also be exposed to the carcasses of poisoned pest animals. Dogs hunting underground may get stuck and starve to death unless dug out. Farm terriers sometimes disappear for days or even weeks and then reappear having eventually escaped from some burrow. Some types of prey are dangerous to the hunting dogs. Pigs, deer and all the large cats can defend themselves and inflict severe injuries on hunting dogs.

A condition know as limber tail caused by damage to the coccygeal muscles has been identified in English Pointers after being worked hard in cold conditions, or held in a small cage for a long period (Steiss *et al.*, 1999), and in Labrador Retrievers after hunting (Wilkins, 1997), swimming (Jeffels, 1997) or being showered in cold water (Hewison, 1997; Stockman, 1997). Hunting dogs, hounds and working dogs are more likely to suffer from leptospirosis than other dogs (Ward *et al.*, 2002).

Dogs used for shotgun hunting of birds, ground game or vermin may be shot accidentally (Keep, 1970. Many injuries caused by shotgun pellets go unnoticed especially if caused by small shot (Keep, 1970) and are often first noticed when the dog is radiographed for other reasons. Lead pellets are rapidly encapsulated in fibrous connective tissue and surgical removal is not necessary unless the pellet is lodged in a joint when the synovial fluid dissolves the lead and causes toxicosis (Grogan & Buchholz, 1981). There is a shift away from lead to steel shot in wildfowl hunting. In future dogs may be peppered with steel rather than lead shot. Steel shot corrodes in tissues and caused severe inflammation (Bartels *et al.*, 1991). If bacterial infection occurs then draining tracts could develop and the condition would be more serious than that caused by lead.

5.4   Draft dog

In a few countries dogs are still used as traction animals. The use of dogs for pulling carts was one of the reasons why welfare organisation became interested in the welfare of dogs in the 19th century and in the UK the use of dogs for transport was banned in 1854 (Hubrecht, 1995). The use of animals for traction is always associated with specific issues such as overloading, long hours of work, too little and too poor food, and physical abuse. These were certainly associated with the use of dogs as draught animals but the practice is now rare and more of a novelty than an economically significant activity.

5.5   Service dogs

Dogs are used by police and military forces around the world for tracking, guarding and identification work. In addition many dogs owned by individuals are involved in search and rescue work. Police dogs have to be physically capable and of a temperament suited to the type of work they do. Up to 70% of dogs bred at the South African Police Breeding Centre were not suited for use as police dogs as they did not meet the standard required for advanced training (Slabbert & Odendaal, 1999). In New Zealand about 40% of dogs bred for police work are successful, but a much smaller percentage of donated dogs become active police dogs (Kyono, 2002). In this study of 74 police dog handlers, many thought that dog and handlers were not well matched (Kyono, 2002). In one sample of 40 police dogs, 24 dogs had had one handler, 13 had 2, 3 had 3 and 1 dog had 4 handlers. This suggests that up to 40% of dogs were mismatched initially (Stafford *et al.*, 2003)

Police dogs working crowd control can be injured by glass and more rarely by petrol bombs, flying bricks or stones. Police dogs can be injured when tracking criminals. In some countries police dogs can only bail up criminals, but with particularly vicious individuals it may be better if the dog is allowed to bite and hold. Police dogs are stabbed, beaten and killed by criminals. Dogs trained to bail up rather than bite and hold are particularly susceptible to attack from criminals. The New Zealand police have had 20 dogs killed on active service since 1956.

Tracking from the scene of a crime is a common practice for police dogs. This can be stressful, particularly at night, when both dog and handler may be in danger. Some very successful police dogs get stressed by this activity and may lose weight and have persistent diarrhoea. They usually recover after a break but need to be managed sensitively. Dogs appear to cope with shift work quite effectively probably due to their short sleep-wake cycles (Adam & Johnson, 1994b)

Drug and bomb detection dogs need to be physically fit, love praise, have a strong hunting instinct, and be both focussed and have sustained attention (Mozdy, 1997). This combination of characteristics makes for a hard-working-dog and it may have to be taught when to work, and when to rest (Luescher, 1993; Rouhi, 1997). Drug and bomb detection dogs sniff up to 300 times a minute when working and this is probably strenuous enough without having to search actively through baggage. Following extreme physical activity there is a decrease in sniffing frequency and an increase in panting with a consequent decrease in explosive detection (Gazit & Terkel, 2003) and moderate physical conditioning will help dogs maintain olfactory acuity (Altom et al., 2003). Dogs that search for drugs or explosives may become stressed and stop working effectively. This may be due to the difficulty of the work, its duration or an inappropriate reinforcing schedule (Luescher, 1993). Drug detection dogs may ingest drugs before their handler can reach them and become poisoned by the drugs (Dumonceaux & Beasley, 1990). In Australia many drug and custom detection dogs work for up to eight years which suggest that the work is not too distressing.

Dogs are also an important tool in the detection of wildlife and other biological materials being smuggled across borders and thus help maintain nations remain free of disease that may affect endemic flora and fauna and commercial livestock, poultry and crops.

Landmine (McLean, 2001) and bomb detection work are obviously dangerous, but drug detection may also place the dog in physically dangerous environments. Trained landmine detection dogs are valuable. They are protected by their handlers, but nevertheless this is a dangerous job for dogs. Nevertheless, mortality rates in mine detecting dogs are very low and in Afghanistan 3 or 4 dogs have been killed in 15 years of operations with up to 140 dogs working 6 days a week (Ian McLean, Geneva International Centre for Humanitarian Demining, personal communication 2004).

Military dogs can be deployed in countries which have diseases and environmental conditions which are foreign to them. In Vietnam the injuries of American military dogs varied depending on the work; 15% of scout dogs were injured severely enough to stop work while only 5% of tracker or sentry dogs were injured (Jennings et al., 1971). In one study of 927 American military dogs the majority died or were euthanased because of diseases associated with advanced age (Moore et al., 2001). The leading causes of death were degenerative joint disease, neoplasia, spinal cord disease, geriatric decline and gastric dilatation-volvulus. The latter is to be expected as these dogs are often large, deep chested animals (Herbold *et al.*, 2002). Military dogs are expensive to produce and consequently receive very good veterinary care. Of 2,500 American military dogs subject to necropsy, 18 died from injuries caused by bullets or bombs, and 17 died from heat stroke (Robinson & Garner, 1973). The average age at death or euthanasia was 10 years and those that died before eight years of age usually died due to heat stroke or a mishap during anaesthesia.

In the past, serious weight loss (10%) was observed in military dogs following short intense tours of action and this was considered due to the constant stress and tension that the dogs experienced (McNamara, 1972). The adrenal glands of military dogs were sometimes five times as large as ordinary dogs. High calorific density diets were recommended to maintain bodyweight (McNamara, 1972) but presumably the high levels of stress continued. About one quarter of military dogs with behavioural problems engage in repetitive behaviour, often spinning in a tight circle (Burghardt, 2003). Stereotypic behaviour like this is probably stress related.

Search and rescue dogs are used in urban and rural disaster and search work. They may have to work after natural disasters such as earthquakes, hurricanes, storms, floods and avalanches and man-made disasters such as explosions, fires and bombings. They may also be used to search for lost hikers and for cadavers. In many of the situations in which search and rescue dogs work there are definite hazards. These may be physical or chemical.

In the search and rescue operation after the bombing in Oklahoma City in the USA in 1995, details of health and injuries sustained by 69 of the 74 dogs used at the site over 491 dog days (Duhaime *et al.*, 1998) were obtained by surveying their handlers. Of the dogs, 46 were search, 14 patrol, 12 explosion detection and two search/patrol dogs. Nineteen dogs (28%) were injured. Footpad injuries were most common (18 of 20 injuries) and 15 (22%) became ill. Search dogs were more likely to be injured during the first two days of work. Older German Shepherd Dogs were especially succeptible to injury (Duhaime *et al.*, 1998). Dogs did not get foot injuries when they wore booties. Many of the illnesses were management related, diarrhoea or vomiting from eating too much or too many treats, but some dogs had irritated eyes and respiratory problems and became hypothermic. These dogs were walking over broken glass, reinforcing iron rods and other slippery and sharp surfaces and overall considering the conditions, the number of injuries was small. When they returned home after the exercise many dogs were tired and stressed. Duhaime *et al.* (1998) recommended that if dogs are used at such sites then they should have cages at the work place and an isolated rest zone which is quiet and where they can rest without being disturbed. They should also be washed frequently and wear booties.

When working in urban disaster sites dogs are often exposed to a range of toxic chemicals (Gwaltney-Brant *et al.*, 2003; Murphy *et al.*, 2003) and may absorb them through the skin or nose. If they are doing a lot of scent work then they need to have their noses wiped regularly and to be bathed and rinsed often and kept well hydrated (Wismer *et al.*, 2003). It may be advisable to provide the dogs with booties and goggles to reduce intoxication (Wisner *et al.*, 2003).

The breeding and selection of dogs for service work has been investigated so as to maximise the success of breeding and training programmes. The success of these means fewer animals are rejected and have to be re-homed or euthanased. Heritability for success or appropriate temperament is reasonably high for service work (Mackenzie *et al.*, 1985; Wilsson & Sundgren, 1997) which suggests that breeding for success is relatively easy. However developing behaviour tests on young dogs to identify those suited for service work have not been so successful. Nevertheless, Slabbert and Odendaal (1999) found that the retrieval behaviour of eight-week old pups and aggression at nine months could predict adult police dog efficiency.

5.6   Cattle and Sheep dogs

When sheep and goats were domesticated, dogs were initially used for guarding flocks. This practice continues today with specific breeds being developed for specific guarding characteristics (Green & Woodruff, 1988). This work entails defending flocks from predators and thieves. These dogs have to be willing to attack bears, wolves, coyotes, other dogs and thieving humans. When dogs are used to protect livestock, especially sheep, from predators (Linhart *et al.*, 1997; Hansen *et al.*, 2002) then the size and type of predator and the number and ability of the guard dogs will determine the level of danger they experience. Bears, large cats and wolves are able to kill dogs. Coyotes are more likely to try and avoid dogs.

When the size of the herds and flocks owned by humans increased greatly dogs were selected and trained to drive animals in addition to being guard dogs. There have been many types of driving dogs. In New Zealand two types of sheep dogs were bred in the last century for helping shepherds manage the large sheep flocks common in that country. One type is the Huntaway, a dog bred to bark and drive sheep from behind. The other type is called a Heading Dog or an Eye Dog and it works to direct sheep and it is similar to the Border Collie. Severe selection would have been fundamental to the production of these two types of dogs and many unsatisfactory dogs would have been killed in the process. The working of sheep is generally safe but injuries do occur (Table 3) with subluxation and hyperextension of the carpus being common (Walker, 1997a, b).

Table 3. The type and cause of serious injuries sustained by 69 working sheep and cattle dogs in New Zealand.

Breed	Cause	Injury Type
31 Heading Dogs	Falls 2	Fractures 36
29 Huntaways	Fence/gate 6	Joint luxations or
6 Border Collies	Livestock17	Ligament instability 35
3 Bearded Collies	Vehicle 19	Muscular or tendon
	Unknown 25	problem 6

Adapted from Walker, 1997a

Dogs are used to drive and direct cattle by barking, snapping at the heels, and sometimes grabbing the nose. Thus, while Corgis are heel snappers many dogs used on wilder cattle snap at the animals' heels, but also grab the face to turn cattle and get them to move. This is dangerous work, and cattle dogs when young and naïve, or old and slow, are kicked, gored, head butted and stamped on by cattle.

5.7   Fighting dogs

In some parts of the world dogs are used for baiting bears, bulls or pigs, or for fighting. These activities have obvious significant effects on the health and wellbeing of the animals involved. Fighting dogs inflict serious injuries on each other. Specific breeds and strains of dogs are bred for fighting, but dogs of other breeds may be used as sparring partners during training and they may be injured seriously. Dog fighting is legal in several states in the USA and in Japan. In the UK, dog fighting is illegal but it was considered well established in 1988 by Wilkins *et al.* (1988). The Tosa (Japanese Fighting Dog) breed and dog fighting appears to be declining in Japan but dog fighting appears to be booming in the USA with both impromptu street fights and organised pit fighting. These impromptu street fights usually end before serious injuries are sustained, but in organised pit fighting death or serious injuries occur.

Not surprisingly, there is very little information about the injuries caused by dog fighting in the veterinary literature, but two reports, one from the USA and one from the UK, give an idea of the injuries involved (Clifford *et al.*, 1983; Hnatkiwskyj, 1985). Clifford *et al.* (1983) examined 32 dogs (American Pit Bull Terriers, American Staffordshire Terriers, Staffordshire Bull Terriers, Mixed breeds) that had been removed from the homes of dog fighters and detailed their injuries (Table 4). Hnatkiwskyj (1985) described a Staffordshire Bull Terrier after a fight with the entire skin and muscles covering its lower jaw ripped off and its mandible exposed for half its length. This dog was treated and during the days in the veterinary surgery the dog showed no concern for her wound and ate vigorously. Fighting dogs appear to have a very high pain tolerance (Clifford *et al.*, 1983), which is presumably necessary to keep fighting.

Table 4. Injuries of fighting dogs.

	Adult Males (n = 9)	Adult Females (n = 17)
Wounds	1	2
Abscesses	1	2
Scars		
Ears	4	9
Muzzle	4	9
Throat	1	5
Frontlimbs	5	13
Hindlimbs	6	10

Adapted from Clifford *et al.*, 1983

The American Pit Bull Terrier, which was bred to fight in a pit, has gained worldwide notoriety because of the breed's vicious behaviour, and its attacks on humans. It has been banned in some countries and restricted in others. Some strains of this breed may be non-aggressive, but Mason (1991) observed 30 adult Pit Bull Terriers in a shelter over a period of several months and was shocked by their agility, aggressiveness and destructiveness. A tyre would be destroyed by one dog in a day and they had to be kept out of sight of each other. The decision to destroy all of these terriers in the UK caused quite a furore but veterinarians who wrote to the Veterinary Record appeared generally in favour of destruction (Flower, 1991). Fighting dogs are bred to fight, to be game for fighting and proponents of the activity argue that because they are bred to fight and apparently are game for it then it is acceptable. This is a spurious argument and does nothing to offset the injuries which occur during the training and fighting of these dogs.

5.8   Racing dogs

Many types of dogs, including several gazehound breeds, sled dogs (Chapter 8) and terriers are raced for sport, but the Greyhound is the ultimate canine athlete having being bred for field coursing and track racing for many years (Chapter 8). There are several aspects of greyhound racing which cause concern: the wastage of dogs that never race, injury during training and racing, the length of their racing career and what happens to Greyhounds when their racing career is over. These are discussed in Chapter 8.

5.9   Assistance dogs

Dogs are used to help people who are visually impaired, have poor hearing or are disabled in other ways. Dogs are probably well suited to this type of work, being focussed on human behaviour (Pongracz *et al.*, 2001) and capable of cooperating with humans (Naderi *et al.*, 2001). Nevertheless, cooperating with people who have sensory, physical and/or mental deficiencies may be a stressful experience. Guide dogs for the blind may have a difficult and stressful job. The success rate of matching dogs to their handler was found to be about 74% in one New Zealand study of 118 dogs (Lloyd, 2004) and Nicholson *et al.* (1995) reported that 15 of 59 dogs (25% were withdrawn from work. Some dogs considered by their handlers to been a mismatch were not returned to the guide dogs providers but kept and worked (Lloyd, 2004). The welfare of these dogs may have been compromised if they were used by people who found aspects of their behaviour difficult. Mismatched dogs are usually given to a second handler. Of 20 dogs identified by Lloyd (2004) as being re-matched 5 failed completely and of the remaining 15, four were on their third handler.

In a survey of the success rate of guide dogs, more handlers considered that they had not been matched successfully with their second guide dog (31%) than their first dog (20%) or third dog (13%) (Lloyd, 2004). More second dogs were returned for non-work related issues than either first or third dogs. Guide dogs are about 18 months to two years of age when allocated to a handler and in Lloyd's (2004) study, 79 dogs had a working life of on average 4.7 years. The working life of guide dogs is influenced by their deteriorating health and they becoming too slow for their handlers.

The temperament of individual guide dogs influenced their heart rate during training and some were apparently less capable of damping down large changes in heart rate than others (Vincent & Leahy, 1997). The Labrador Retriever has been the guide dog of choice for many years. This is possibly due to their high trainability and physical size, however their low reactivity (Hart & Hart, 1985b) is also important. Guide dogs spend a lot of time doing nothing and then have to take responsibility for guiding, a breed with an easy relaxed nature is necessary for this type of lifestyle.

Individual dogs may be matched with a person with serious physical and/or mental problems and this may make the work of the dog very difficult. However guide dogs appear to be able to change their role from that of leader to that of being led, depending on circumstances. In some guide dog–human dyads, the dog initiated most actions but in others it was the reverse and in none did any party initiate more than 80% of actions (Naderi *et al.*, 2001). This allows the dyad to develop new types of action and may reduce the stress on the dog and its handler as both work together in a cooperative manner.

Dogs trained to assist people with physical disabilities have surprisingly physically active roles and Lane *et al.* (1998) were concerned about the use of breeds susceptible to hip dysplasia for this work. The attitude of the handlers of these dogs was influenced by whether getting the dog had been their idea or not. If the decision to get a dog was not that of the handler the dog was likely to be regarded as no more than a work animal and such handlers tended to agree that it might be more trouble that it is worth. However, there was no evidence that the welfare of dogs of these handlers were any less well treated than those of handlers who chose to acquire a dog (Lane *et al.*, 1998).

5.10   Therapy dogs

Dogs may assist in the diagnosis of some diseases. They are used to help people in hospices and hospitals. The work itself should not be particularly stressful and their welfare is discussed in the chapter dealing with companion dogs (Chapter 12)

5.11   Entertainment dogs

Dogs are commonly used in advertisements, traditionally in circuses and occasionally in zoos. In circuses, advertising and the stage, dogs are often trained to do tricks and while some consider these acts to be demeaning it is unlikely that the dogs think of them as such. Good entertainment dogs usually appear to enjoy their work and provided it is not damaging there are probably no major concerns for the dogs' welfare.

5.12   Laboratory dogs

Dogs are used in research laboratories for a variety of purposes. They are used in models of human disease and for research relating to canine health and nutrition. The use of any animal in a laboratory or testing laboratory causes concern for their welfare, but concern for the dog has resulted in a reduction in dog usage by scientists in many countries over the last decade (see Chapter 9). Some dogs are rehomed after careers as laboratory dogs.

5.13   Sport dogs

Dogs are used for showing, agility and obedience competitions and many types of field trials.

Showing dogs is a very popular sport and forms the backbone to the breeding of most pedigree dogs (Chapter 3). Breed standards are discussed in chapter 3 but the process of showing and preparation for showing are discussed here. To succeed in the show-ring dogs have to be presented as expected for the breed and behave as is appropriate. All dogs have to be clean and for some their hair has to be dressed in an appropriate style. Dogs in the show-ring have to be non-aggressive and confident if they are to succeed. These are all good for the welfare of the dogs being shown. However, there are specific breeds that have to be modified physically to enter the show-ring. These modifications include ear cropping, tail docking and dew claw removal.

Ear cropping is banned in many countries but in others it is routinely carried out on fighting, show, guard and pet dogs. In The Complete Dog Book of the American Kennel Club (Anonymous, 1998a) the ears of the Doberman Pinscher, Boxer and Miniature Schnauzer must be cropped, while those of the Giant and Standard Schnauzer, Great Dane, Bouvier des Flandres, Miniature Pinscher, Affenpinscher, Manchester Terrier and Boston Terrier may be cropped or not. Ear cropping is not carried out in some countries (UK, New Zealand, Ireland). It is carried out when the puppies are weeks or months of age (Table 5). The procedure is certainly painful and there are the attendant risks of anaesthesia, haemorrhage and infection (Nolen, 1999). Many object to ear cropping as it is carried out purely for cosmetic purposes (Anonymous, 2003d) but others argue that if a particular society desires such a procedure then veterinarians should carry out the procedure (Stone, 2000) rather than have it done by anyone else.

Table 5. The age when ear cropping should be carried out and the amount of *ear* remaining after surgery.

Breed	Age	Amount of ear remaining
Schnauzer	10 weeks	2/3 remaining
Boxer	9-10 weeks	2/3 to ¾ remaining
Doberman Pinscher	8-9 weeks	¾ remaining
Great Dane	9 weeks	¾ remaining
Boston Terrier	4-6 months	Full trim

Adapted from Slatter (1993)

Tail docking has been a major source of disagreement between kennel clubs and animal welfare organisations (Bennett & Perini, 2003) for the last twenty years or so. It is carried out worldwide with only a few countries where it is not allowed. The welfare organisations and many veterinarians (Morton, 1992; McCreath, 1993; Wansbrough, 1996) argue that it is painful and unnecessary and removes a major means of communication whilst the kennel clubs and many veterinarians argue that it is of little significance, necessary, virtually or completely painless, traditional and that dogs with long tail can damage them easily (Dean, 1990; Webster, 1992; Mercer, 1992; Collins, 1993; Brown, 1998a; Warman, 2004). Many breeds of dogs are tail docked (Table 6) and the procedure is usually carried out in the first few days of life by surgical removal using scissors or scalpel sometimes followed by a suture (Slatter, 1993), or by an elastic band. In the past dogs were docked for spurious reasons, such as to prevent rabies, to prevent tail injuries while fighting, and in some countries to avoid paying tax as only the dogs with a long tail were considered for tax purposes (Wansbrough, 1996).

Tail docking had no prophylactic effect in reducing tail injuries in the study of Darke *et al.* (1985). There was an association between docking and acquired urinary incontinence due to incompetence of the urethral sphincter mechanism, but this may relate to a breed predisposition rather than be caused by docking *per se* (Holt & Thrusfield, 1993, 1997). Brachycephalic breeds that have their tails docked may be predisposed to perineal hernia (Burrows & Ellison, 1989), which may be due to poor muscle development in docked animals (Canfield, 1986 cited by Wansbrough, 1996) but this has not been confirmed by research.

When puppies were tail docked using scissors they shrieked at the time of amputation but settled down to sleep after about 3 minutes (Noonan *et al.*, 1996). This suggests that the procedure is painful but the pain is short lived. The pain caused by docking in puppies cannot be compared to that in lambs as the former is much less developed when docked than the latter. The development of neuromas (Gross & Carr, 1990) and adhesions (Carr, 1979) may be painful, and dogs may lick and bite at their tails which have to be shortened again (Scott *et al.*, 1995). Dogs may develop a phantom tail but this has not been determined. Pain in newborn and premature children is now assumed to occur. Schuster and Lenard (1990) suggest that those who carry out potentially painful operations without analgesia on neonates should have to prove that the latter do not feel pain. To carry out such surgery is to go against everyday experience and scientific knowledge. Boys who had been circumcised displayed more pain behaviours at four and six months of age when vaccinated than uncircumcised boys (Taddio *et al.*, 1995) so it is possible that docking may have long term effect on pain perception in dogs. Wansbrough (1996) was concerned about the development of pathological pain in puppies following docking, but there is no evidence of this happening. There are likely to be differences in the pain experienced during and after short docking (eg. the rottweiler which is left with one or two coccygeal vertebrae) and long docking (eg. the wirehaired pointing griffon which is left with two thirds of its tail), but this has not been investigated.

Table 6. Breeds of dogs that are routinely or sometimes docked in the USA (Anonymous, 1998a).

Sporting Breeds	12 of 24 breeds docked
German Shorthaired Pointer	docked leaving 40% of its length
German Wirehaired Pointer	docked to 2/5 of its original length
Clumber Spaniel	docked in keeping with the overall proportion of the adult dog
Cocker Spaniel	Docked
English Cocker Spaniel	Docked
English Springer Spaniel	Docked
Field Spaniel	docked to balance the overall dog
Sussex Spaniel	docked from 5-7 inches
Welsh Springer Spaniel	Docked
Visla	tail should reach back of stifle joint
Weimaraner	measure 6 inches at maturity
Wirehaired Pointing Griffon	docked by 1/3 ½
Hound breeds	none of 22 breeds docked
Working breeds	5 of 20 breeds docked
Boxer	Docked
Doberman Pinscher	docked at approximately second joint
Giant Schnauzer	docked to the second or not more than the third joint
Rottweiler	docked short or close to the body leaving 1 or 2 vertebrae
Standard Schnauzer	docked to not less than 1 inch or more than 2 inches
Terrier breeds	10 of 25 breeds docked
Australian Terrier	docked in balance with the overall dog
Wire Fox Terrier	docked a ¾ dock is about right
Irish Terrier	docked taking off about 1/4
Lakeland Terrier	docked
Minature Schnauzer	docked
Norfolk Terrier	docked (medium)
Norwich Terrier	docked (medium)
Sealyham Terrier	docked
Soft-coated Wheaten Terrier	docked
Welsh Terrier	docked to a length approximately level with the occiput
Toy breeds	6 of 19 breeds docked
Affenpinscher	docked or left natural
Brussels Griffon	docked to about 1/3
Cavalier King Charles Spaniel	docking is optional, no more than 1/3 to be removed
English Toy Spaniel	docked to 2 to 4 inches in length
Miniature Pinscher	docked in proportion to the size of the dog
Silky Terrier	Docked
Non-sporting breeds	none of 16 breeds docked
Herding breeds	4 of 17 breeds docked
Australian Shepherd	docked or naturally bobbed, not to exceed 4 inches
Bouvier des Flandres	docked leaving 2 or 3 vertebrae
Old English Sheepdog	docked close to the body
Pembroke Welsh Corgi	docked as short as possible without being indented
Miscellaneous class	2 of 5 breeds docked
Jack Russell Terrier	docked to the tip is approximately level to the skull
Spinone Italiano	docked

A total of 39 breeds docked out of 148 breeds

Table 7. Criteria to test the necessity of elective surgery on the dog.

Is there evidence that leaving the dog intact predisposes them to harm?

Is there evidence that surgery is in the best interests of the dog and is beneficial to it?

Would the harm or benefit occur in a significant proportion of dogs and therefore justify the surgery on a particular breed?

Does the surgery cause greater harm than the damage one is trying to prevent?

Is there another way with fewer or no adverse effects to achieve the same end?

Does the increase in value as a result of surgery justify the damage done?

Adapted from Wansbrough (1996) and Morton (1992)

Tail docking has been banned in several countries (Israel, Finland, Switzerland, Germany, Sweden, Norway, Australia) and will be banned in more countries soon. The prohibition of docking appears to have occurred primarily because it is a cosmetic and painful procedure. However there has been little effort to develop pain control protocols in the young pup. Local anaesthetic can be used to alleviate the pain caused by docking but Wansbrough (1996) was concerned that a 2% solution of lignocaine might be toxic to young pups. General anaesthesia in very young animals may be dangerous. If analgesia was used in the young pup before, during and after docking then the pain-related argument against docking would be weakened. However, the possible side effects such as haemorrhage and infection, and potential long term effects will remain arguments against the procedure. Moreover, the removal of the tail may influence the ability of dogs to communicate effectively with conspecifics and it also makes it more difficult for humans to understand them.

Morton (1992) developed six criteria to test the necessity of docking (Table 7), and Wansbrough (1996) found that the general reasons advanced for docking dogs tails did not satisfy these criteria and concluded that it cannot be justified. However, the docking of specific breeds engaged in work may meet the criteria, thus in Germany docking is banned, unless it is absolutely necessary for hunting, and then must be carried out by a veterinarian. This might satisfy those who believe that working English Springer Spaniels require the last third of the tail removed to prevent injury in the hunting field (Webster, 1992; Neal, 1992), but there is no published evidence to support their concern about injury to this breed during hunting.

Dew claws are the first digit of the limbs. In hunting dogs the dewclaws are removed from the hind legs to prevent damage, and in toy breeds they are removed from front and hind limbs to facilitate clipping (Bojrab *et al.*, 1983). They are usually removed at 3-5 days of age (Slatter, 1993) when haemorrhage is usually small. They are cut off with scissors or a scalpel blade and sutures are not usually placed. In many breeds it is recommended that the dewclaws be removed (Anonymous, 1998a) but in some breeds such as the Kerry Blue Terrier, dogs are disqualified if they have dew claw on hind limbs. If dewclaws are damaged on an adult dog then the dewclaws may have to be removed surgically under general anaesthesia.

Agility, flyball and obedience activities are often recommended to keep a dog busy and to develop the dog-human bond. Dogs may find agility, flyball and obedience competitions stressful but whether these activities are damaging or not is unclear. It is likely that successful dogs find them exciting and some of those that do not succeed find them too stressful. Linda Beer (2003) suggest that frustration is an important source of stress and that it is used in training dogs for flyball to increase speed. If the dog's frustration level is allowed to escalate the dog may become aggressive towards its handler or other dogs. She felt that, if possible, dogs awaiting their turn to compete should be held out of sight of the competing animal. There is a need for more data on the stress experienced by competing dogs and how success and failure in these sorts of competitions affects different dogs.

5.14   Companions

The close human dog relationship probably developed during the later stages of domestication when the wolf/dog was willing to come into contact with human beings. This change in behaviour was observed during Belyaev's study of fox domestication when foxes approached humans and became willing to physically contact them (Trut, 1999). The human need for heat probably accelerated the closeness of the relationship and dogs were used as bed warmers by Aborigine people in Australia (Meggitt, 1965). The shift from wary associate to companion was a driving force in the human-dog relationship and generally this relationship has continued to benefit both species. However Meggitt (1965) observed that wild dingoes were in better condition than tame dingoes, as the latter were only fed bones and then left to fend for themselves.

This type of companionship is common between humans and other species including cattle and pigs. Dogs may be able to read human intentions effectively (Hare *et al.*, 2002) but humans may not be so sensitive to what dogs are intimating. However, humans can recognise what dogs mean by their different types of barking (Douglas, 2004). Many dog owners regard their dogs as companions and work mates, but these dogs sleep outside and are fed and managed separately from humans and often from their pet dogs. Companion does not mean family member. The statement that people regard their dog as a family member needs to be interpreted cautiously and the number of dog behaviour problem books published in the last two decades suggests that there are serious problems in the human-dog relationship. Human expectations may be greater than what is possible.

Domestication may be a process of reduction in ability to cope quickly with the dangers and stresses of the natural world, a process which results in an animal with a tolerance of stress, a docile nature and a lack of fear (Hemmer, 1990). This process is ongoing and may be fundamental to the long-term success of the dog as a house dog and companion. The difficulties experienced in post-industrialised societies in the last half century with dog behaviour may have been a result of the movement of the dog off the street, for welfare reasons, and into a much less interesting environment, the house. This movement has not been accompanied by a deliberate selection of animals to cope with such an environment. A minority of breeders may breed from their own house dog but much pedigree dog breeding continues to select for physical characteristics and temperaments which may or may not be suited to the relatively dull life of the modern companion animal (Chapter 12).

6.   CONCLUSION

The domestic dog has lived with humans for thousands of years and has been and remains a useful companion in many of our endeavours. In many of these activities the dog is put in danger and dog fighting is often prohibited because of the damage caused to the dogs. The distress or lack of distress caused by many of the activities that dogs are used in has been poorly investigated, and the health and longevity of working and sport dogs has been poorly defined. There is a need for research into the distress caused to dogs during work and sport, and during the training of dogs for these activities. The dog has a limited capacity to cope with all of our expectations and unless its physical and social requirements are met then its welfare will be compromised. Compromise may occur when the animal's physical needs are met but its social and psychological needs are not.

Chapter 2

FREE LIVING DOGS

Abstract: The majority of dogs are free-ranging and may or may not be owned. Many
are appreciated for their sentinel activities and disposal of garbage but there
are also fears in many countries of zoonotic diseases especially rabies. The
nutrition and health of these dogs is poor and their lives short and possibly
brutal. In most wealthy developed countries free-ranging dogs are uncommon
as dog control legislation forces owners to identify and restrict their dogs and
dogs found roaming are caught, held and either reclaimed, re-homed or
euthanased. The means to control free-ranging dog populations are well
known, but the political will or the financial capability to do so is often not
available. Free-ranging dogs become a political issue when zoonoses or dog-
attacks become important and then dog control programmes may be initiated.
These are often under funded, incomplete and short lived. Legislation, animal
control and education are essential to dog control. De-sexing and killing are
standard tools in dog population control but education, registration, and habitat
modification are also important. The methods used to kill dogs vary depending
on philosophy, funding and the availability of drugs and trained personnel.
The method used to control free-ranging dogs varies, depending on the
philosophy of the community. In some countries where killing is not
acceptable, de-sexing and habitat adjustment are necessary to control free-
ranging dogs. Dogs are highly fecund and if food is available the removal of
dogs from a location will be rapidly followed by the immigration of dogs from
other areas. The welfare of free-ranging dogs is the most significant welfare
issue of dogs and assistance from animal welfare organisations in the
developed countries to animal welfare organisations and veterinarians in
developing countries may be the most effective way to improve dog welfare
worldwide.

1. INTRODUCTION

The world population of domestic dogs has been estimated at around 500 million (Matter & Daniels, 2000) and by some estimates, about 75% of these animals are free-ranging. These figures are approximations as it is difficult to determine the population of dogs. This is true even in countries where the majority are owned as pets or working animals (Patronek & Rowan, 1995) and it is even more difficult to determine what percentage of dogs are owned and free-ranging, or not owned and free-ranging or wild. There is a problem with dog population control in many countries (Leney & Remfry, 2000) and this has serious public health and dog welfare implications. It is believed that the problem with free-ranging dogs, dog-related zoonoses (Cleaveland *et al.*, 2003) and dog attacks is increasing as is the threat posed by dogs to wildlife in some countries (Pain, 1997; Butler *et al.*, 2004)

A dog may be a beloved pet, a valuable working animal or an incidental animal living with humans without any major function. They may also be unwanted but continue to live close to humans feeding off garbage, or become wild, living distant to humans and without any direct dependence on humans (Boitani *et al.*, 1995). In simple terms, dogs may be owned valued and restrained, owned but free-ranging, not-owned and free-ranging, feral or wild (see Chapter 1, Table 2). The term 'stray' refers to owned dogs living with humans, but allowed to be free ranging, or to dogs living close to humans, dependent upon human refuse or handouts for food but not owned. Wild dogs are domestic dogs that have returned to the wild permanently. They may live away from human habitation but often live on city dumps.

In discussing the welfare of free-ranging or wild dogs it is necessary to look at the background to human concerns about these animals, and the efforts made by humans to control the population of dogs worldwide. The human interest in free-ranging dogs ranges from a fear of rabies and other zoonoses, worry for the safety of livestock or wildlife or a genuine concern for the welfare of these dogs. The type of interest will depend on the lives of the people concerned, their own welfare and quality of life, religious beliefs and wealth. It is foolish to expect people who are struggling to feed and rear their children to be too concerned with the survival of a mangy scavenging dog or to expect people to whom the dog is by religious conviction a dirty animal to worry about the mortality rate of newborn pups. However it may be that the concern for human health, especially the drive to reduce the incidence of rabies, may also improve the welfare of dogs in many countries by reducing their numbers and controlling unwanted pup production. Many free-ranging dog become victims of road traffic accidents. They may be beaten severely or killed by humans or other dogs if they behave in what is considered to be an inappropriate manner.

Dog population control is an important aspect of public health programmes in many countries. The control of free-ranging dogs is also important for livestock protection and for conservation reasons in some countries. In some societies the presence of stray or wild dogs is unacceptable for philosophical and/or aesthetic reasons and in these societies dog control is carried out not because of public health issues but for animal welfare reasons. In this chapter the welfare of free-ranging dogs will be discussed in relation to why they are considered a nuisance and how they are controlled.

2.   DOGS AS NUISANCE

The dog is a barking, scavenging, territorial, sexually promiscuous, social, pack protecting animal with a propensity to defaecate or urinate just about anywhere. These behavioural characteristics do not lead to neighbourhood harmony. A brief litany of the anti-social behaviours of dogs include chasing cars and bicycles, causing car accidents, attacking livestock and other companion animals, damaging property, producing unwanted pups and attacking humans (Murray, 1993). The positive effects of dog ownership are relevant to the individual (Hart, 1995) rather than society and dog behaviour is a common cause of disagreements amongst neighbours. In Melbourne, Australia, the most common category of neighbourhood dispute related to a neighbour's animals, usually a dog or cat (Technisearch, 1990), and in Queensland, Australia, municipal authorities ranked dog problems as their second greatest management problem, following rates collection (Murray, 1993). In the 1970s in the USA 60% of mayors reported that animal problems led the list of complaints from the public (Katz, 1976).

Table 1. Dogs caught and handled by the police and dog wardens in the UK and dog control personnel in Japan.

	1988	1996
UK*		
Dogs caught as strays	240,000	139,000
Reclaimed	60,000	46,500
Dogs adopted or sent to shelter	90,000	51,600
Dogs euthanased	90,000	16,200
Japan**	1989	2000
Dogs caught as strays	297,454	126,570
Reclaimed	12,210	15,004
Euthanased	285,224	111,566

*Adapted from Leney and Remfry (2000)
** Ministry of Health, Japan

The political recognition that dogs can be a nuisance and a cause of zoonotic diseases led to the establishment of public dog control programmes in cities in North America and Europe in the 19th century, programmes which continue to this day. That these programmes continue suggest that there are ongoing problems in the human–dog relationship and urban dog management. The number of dogs picked up off the street in the UK and in Japan has decreased significantly over the last decade (Table 1) and this reflects an improvement in dog control measures plus an increased recognition by the general public of the need for stricter dog control. The establishment of stricter dog control legislation in many developed countries, such as restricting dog breeding and controlling ownership makes dog ownership more difficult and less attractive which will reduce the social problems caused by dogs. If dog ownership becomes more difficult and dog numbers decline then hopefully the welfare of individual animals in developed countries will improve as only determined and dedicated owners will have dogs as pets.

3.   ZOONOTIC DISEASES

Dogs are implicated in the transmission of several important infectious and parasitic diseases of humans including rabies, hydatids disease, visceral larval migrans, and visceral leishmaniasis (Calum *et al.*, 2000). Some of these diseases, especially rabies and leishmaniasis, have major implications for the welfare of dogs in that they are significant diseases of dogs *per se*. The methods of controlling these diseases directly through vaccination or treatment programmes, or indirectly through dog population control may or may not be humane.

World Health Organisation data show that each year more than 50,000 people die from rabies, of which 30 to 50% are children under 15 years of age. Rabies is a very unpleasant disease of the nervous system, which cannot be treated effectively once clinical signs occur. The main route of rabies transmission is through the bite of a rabid dog. Two and a half billion people are at risk of contracting rabies in over 100 countries reporting the disease (WHO data cited by Haupt 1999). The vast majority of rabies cases occur in countries with large free-ranging dog populations, and mortality figures range from 0.001 to 18 per 100,000 people in the USA and Ethiopia respectively (Haupt, 1999). In 1992 about 6.5 million people were vaccinated for rabies after exposure to rabid dogs. Of these, 5 million were in China and half a million in India (WHO, 1994). On a local scale, in the city of Maracaibo in Venezuela there were 207 cases of canine rabies in 1985. Most of these occurred in neighbourhoods with a high dog population, a high proportion of free-ranging dogs and low vaccination rates (Malaga *et al.*, 1992). Some countries are free from canine rabies (e.g. New Zealand, Australia, Ireland, UK), because of strict dog control measures and effective quarantine programmes. Many other countries have rabies under control using vaccination and dog control programmes.

The control of rabies in dogs is a significant public health activity in many countries. It is expensive, and many countries do not vaccinate the minimum percentage of dogs required for rabies control (Meltzer & Rupprecht, 1998). Rabies control is carried out by encouraging owners to vaccinate their dogs or by making it compulsory. In many countries government veterinary services carry out vaccination programmes. WHO figures for 1995 show that more than 34 million dogs were vaccinated against rabies but these figures do not include data from some countries and so is a very low and conservative number. This number was considered to be 41% of the dogs in the countries surveyed. The percentage of dogs vaccinated ranged from 1 to 120% with many countries below the 60 to 70% necessary to prevent the transmission of rabies in dogs (Meltzer & Rupprecht, 1998; Cleaveland *et al.*, 2003).

Rabies vaccination is often accompanied by dog population control programmes. In many countries these involve killing dogs. This is more expensive than vaccination (Meltzer & Rupprecht, 1998) but may have other positive effects with regard to dog attacks and other diseases. Compulsory dog registration, public education, habitat modification, de-sexing, the capture and destruction of free-ranging dogs plus vaccination are standard elements in successful programmes to control rabies in dogs, and all have implications for the welfare of dogs. In general fewer dogs are killed than are vaccinated. From 1984 to 1994 in Sichuan Province in China more than 1.4 million dogs were killed each year, while 26 million dogs were vaccinated against rabies during these 11 years, and the incidence of human rabies dropped dramatically. In Chile, 196,000 dogs were killed over two years in the 1990s, which was about half the number of dogs vaccinated against rabies (Meslin *et al.*, 2000).

Hydatid tapeworms (*Echinococcus granulosus*) are a dangerous parasitic infection of humans, particularly in sheep farming communities. The cysts caused by the parasite in humans can become very large and dangerous, and people can die if the cysts are ruptured accidentally. The tapeworm is of little significance to the health of the infested dog. A standard method of eliminating the disease is to treat all dogs for the tapeworm at regular intervals. Enforced regular treatment as occurred in New Zealand and Iceland (Meslin *et al.*, 2000), may have had other effects as dogs had to be brought to central locations for treatment. Dogs that were unwanted or serving no purpose would probably be killed rather than be brought for treatment.

Visceral leishmaniasis (VL) is a problem in some countries associated with dog ownership (Gavgani *et al.*, 2002a). The disease, which is usually found in children, causes intermittent mild fever, hepatosplenomegaly and anaemia and can be fatal if not treated. In dogs the disease often starts as a chancre on the ear or nose and the parasites infect the lymph nodes and bone marrow. The disease may progress slowly, or rapidly cause death. In Brazil, dogs that are serologically positive for VL are killed, and this, combined with the treatment of human cases and control of insect vectors, has resulted in the eradication of the disease in part of the country (Palatnik-de-Sousa *et al.*, 2001). The treatment of infected dogs was considered prohibitively expensive by Palatnik-de-Sousa *et al.* (2001). In Iran, putting insecticide impregnated collars on dogs reduced the risk of VL in children but did not eradicate it, probably because transmission continues in sylvatic reservoirs (jackals and foxes) (Gavgani *et al.*, 2002b).

Visceral larval migrans is a disease syndrome usually found in children. It is caused by the roundworm, *Toxacara canis,* the common internal parasite of dogs (Overgaauw & van Knapen, 2000). Visceral larval migrans may result in fever, malaise, abdominal pain, poor sight and blindness in children. Control of this disease involves regular anthelmintic treatment of dogs, especially pregnant and lactating bitches, and pups. Reducing the number of stray dogs and preventing dogs from contaminating areas frequented by children such as school grounds and sand pits is also effective (Overgaauw & van Knapen, 2000). Restricting areas where owned dogs can run loose may have a significant effect on available exercise areas. In some countries dogs have to be always on a leash in public or may be allowed to run free only in limited and defined areas. In many developed countries owners are not allowed to exercise their dogs in school yards and playing fields because of the danger of visceral larval migrans. In countries with large populations of free-ranging dogs a great change in dog management and a sizable reduction in the population of free-ranging dogs would be required to reduce the danger of this disease.

Injuries caused by dogs are another major public health concern in many countries. In the USA, dog bite injuries were considered a serious public health problem by Hunthausen (1997) with 1 to 3 million dog bites, and about 18 human deaths each year due to dog attacks. The victims of fatal attacks were usually either very young or very old. In the USA un-owned free-ranging dogs are unlikely to be a significant public health nuisance (Beck, 2000). There are 100,000 free-ranging dogs in Bangkok, Thailand alone (Ratanakorn, 2000), and the people, mostly Buddist, support the feeding and protection of these dogs. Dog bites account for 5.3% of all injuries seen at a teaching hospital in Bangkok. The majority of these are inflicted by free-ranging dogs and are apparently unprovoked (Bhanganada *et al.*, 1993). The danger of injury from dog bites is apparently increasing in some South American countries. This increase may be due to a change in the type of dogs living on the streets. However, without doubt, the majority of dog bites in developed countries are from owned pet dogs rather than free-ranging dogs, although bites from the latter are more likely to be treated medically because of the fear of disease. Dog bites can be serious, but death due to bites from free-ranging dogs is probably rare worldwide (Beck, 2000).

It is possible that the dangers from zoonoses which are transmitted to humans from dogs, will result in more determined efforts by countries with large free-ranging dog populations to reduce the numbers of these animals. This activity would reduce the number of underfed and diseased free-ranging dogs and perhaps allow a better life for those that continue to run free.

4.   DOGS AND OTHER ANIMALS

Domestic dogs may attack and kill livestock, other companion animals and other dogs. They may be important predators of livestock (Schaefer *et al.*, 1981; Blair & Townsend, 1983; Fleming & Korn, 1989) and the killing of cats by dogs is a common source of anguish. In one study of over 1000 incidents of dog attacks on livestock in semi-rural Western Australia in 1989-1990, Jennens (Garth Jennens, personal communication) found that many dog owners allowed their dogs to wander unsupervised and most thought that their dogs were too timid to attack stock. In that study Jennens found a 25 to 30% annual turnover of dogs caused, he believed, by the inability of owners to manage their dogs. The dogs, identified by Jennens, as having chased and killed sheep, were usually friendly house dogs. They chased sheep either alone or in small groups of dogs of many different breeds. Dogs may be encouraged to chase sheep if in the company of a chasing dog (Christiansen *et al.*, 2001b) and some breeds are more likely to chase sheep than others (Christiansen *et al.*, 2001a). In most rural societies dogs that chase and kill sheep are killed if caught in the act.

Dogs are a major threat to wildlife. They may spread disease to other canidae and some felids and may predate on wildlife (Pain, 1997; Scott, 1988; Lever, 1985). In 1994, distemper from dogs killed one third of the lions (*Panthera leo*) in Tanzania's Serengeti National Park. It was suggested that it spread from dogs to the lions via spotted Hyenas (*Crocuta crocuta*) which often mix with lions at a kill. Distemper is also a threat to wild canids including those species that are endangered such as the Ethiopian wolf (Pain, 1997). The presence of dogs within a wildlife research area in Zimbabwe was considered to be a potential source of canine disease for jackals and spotted Hyena (Butler *et al.*, 2004). In that study leopards (*Panthera pardus*), lions and spotted hyenas preyed on dogs. They killed about 6% of the dog population in 1993 thus being exposed to canine disease. Dogs may also spread rabies to other canids and Rhodes *et al.* (1998) argued that as the side striped jackal (*Canis adustus*) population was too low to maintain rabies it had to be regularly introduced by rabid dogs. Predation by dogs on species vulnerable to extinction has been a problem in Australia (Paltridge, 2002), and in New Zealand dogs regularly kill Kiwi.

5.   FREE-RANGING DOG POPULATIONS

The dog population of many countries is usually an approximation. Populations are based on dog registrations in some countries (Hart *et al.*, 1998), but a percentage of owned dogs will not be registered and many free-ranging dogs may be not included anyway. This problem also occurs when dog populations are determined by surveys of households (Odendaal, 1994; Patronek & Rowan, 1995). In specific localities the dog population may be estimated using distance sampling (Childs *et al.*, 1998) or marking recapture methods, techniques typically used for wildlife population surveys (Matter *et al.*, 2000). The dog population appears to be increasing in many developing countries, for example, Zimbabwe, which has a dog population growth rate of 6.5% (Butler, 2000; Butler *et al.*, 2004). The population of free-ranging dogs usually increases when there is a breakdown of civil order, such as during wars and rebellion. Leney & Remfry (2000) suggested that the dog population is related to the human population within the range of one dog per 6 to 10 people except in some countries in the Middle East (Yemen, Syria, Jordan) where the ratio was lower.

The population of any species depends on available resources, its reproductive success and mortality rate. The large canines are fecund. In African hunting dog packs a single litter is produced each year by the dominant bitch and the mean litter size in Selous Game Reserve was about 8 pups (Creel & Creel, 2002) with more surviving to become yearlings in larger packs. Wolf fecundity varies with litter sizes of 3 to 6 pups quoted for European populations by Pulliainen (1975). Wolves may give birth to and rear more pups to weaning than there is space for. In the wild, many canid pups leave the natal pack and either succeed in joining up with other juveniles and carving out a territory, or die through starvation, injury, predation or disease.

One of the effects of domestication has been to reduce the age of puberty and to increase the number of periods of oestrus per bitch per year. Female domestic dogs can breed when less than 12 months of age and produce up to 2 litters per year depending on breed and individual ability. Free-ranging bitches may breed once (Pal *et al.*, 1998a; Pal, 2003a) or twice (Daniels & Beckoff, 1989b) each year. In Italy litters with an average of 5.5 pups were found by Macdonald & Carr (1995). In rural India, free-ranging bitches breed seasonally during October in the late monsoon and the average litter size is 5.7 pups. The majority of pups died within a few months of birth (Pal, 1998a; Pal, 2003b). High mortality rates in pups was also noted in an urban dog population in West Bengal, with 67% of pups dying before 4 months of age and 82% dead within the first year of life (Pal, 2001). In a rural dog population in Zimbabwe's communal lands there was very high juvenile mortality, with 71.8% of dogs dying before one year of age (Butler, 2000). Similar mortality rates are reflected in the demography of dogs in a rural district of Kenya (Kitala *et al.*, 1993). In one study of feral dogs in Italy only two out of 40 pups survived to adulthood (Boitani, 1995). Communal rearing of pups was seen by Daniels & Beckoff (1989b), which is interesting as in wolf packs only the dominant female breeds (Vanhoof *et al.*, 1993).

Bitches can produce 5 to 10 or more puppies each year, so perhaps hundreds of millions of puppies die amongst free-ranging and wild dogs worldwide each year. These pups may die from a combination of malnutrition, disease, predation and exposure, there being insufficient resources for the bitches to rear and protect their litters. The diseases of pups, such as Parvo Virus Disease and Distemper plus internal parasitism caused by *Toxacara canis* are endemic in free-ranging dogs. Unfortunately little can be done about these diseases in large populations of free-ranging dogs that produce large numbers of pups. The dispersal of pups is probably associated with reproductive behaviour. Pal *et al.* (1998a) found that one third of puppies left the natal territory and most of those that did so were juvenile males that dispersed when the females were in oestrus during the monsoon.

The life expectancy of free-ranging dogs in most situations is short. In a classic study of urban free-ranging dogs in Baltimore, USA by Beck (1973) the life expectancy was 2.3 years. In Zimbabwe's communal lands the life expectancy of a dog was calculated to be 1.1 years (Butler, 2000) but had been 4.6 in the district of Manicaland (Brooks, 1990). Despite this, the population is growing in Zimbabwe. The cause of death of older dogs is likely to be a combination of malnutrition and disease, but con-specific aggression and predation by larger predators may play a part and many may be killed deliberately or accidentally by humans. These pup survival rates and life expectancy figures contrast with those seen for puppies born to dogs owned as pets in wealthier societies. The average age of dogs in rural Zimbabwe and Kenya was 2 to 2.3 and 1.8 years respectively (Brooks, 1990;

Butler, 2000; Kitala *et al.*, 1993) and in the former, 40% were less than one year of age.

The reproductive rate of the free-ranging dogs obviously far exceeds the capacity of their environment to support them. The holding capacity of a particular area is determined principally by food availability (Butcher, 1999; Font, 1987). Dogs are scavengers and their diet will reflect what is available in the environment. The diet of free-ranging dogs varies from urban to rural populations. In a study in Mexico, recently abandoned dogs initially survived by living off garbage in a dump but feral dogs rarely visited dumps. This suggests that feral dogs were utilising other food sources (Daniels & Bekoff, 1989a). This contrasts with feral dogs in Italy that regularly visited dumps (Boitani *et al.*, 1995; Macdonald & Carr, 1995). In rural Zimbabwe the diet of free-ranging but owned dogs was made up of porridge (22%), cow carrion (16%) and human faeces (21%) (Butler & du Toit, 2002). Most of the food (88%) was human-derived, but only 13% was actually fed to the dogs, the bulk of it being scavenged. In a Kenyan study, 95% of dogs were fed household scraps (Kitala *et al.*, 1993) and nearly 20% of these dogs were restricted to the household with 69% of them running free all the time.

6.   DOG POPULATION CONTROL

When resources are not limiting then population control depends on either preventing conception, birth or increasing the mortality rate. Dogs give birth to many unwanted and surplus puppies if some form of contraception is not practised. In rural communities where un-owned and uncontrolled dogs may chase and kill livestock these unwanted dogs are kept under control by preventing bitches from breeding, by killing surplus puppies and by killing un-owned and free-ranging dogs (Leney & Remfry, 2000). Puppies may be killed by exposure, starvation, drowning or by one of many other means. In cities this control of the dog population breaks down as the danger from unwanted animals differs and control leaves the hand of the individual and is transferred to local authorities. City people may leave a litter of puppies on the street and hope that someone will take care of them, thus not accepting their responsibility for puppies they produced (Hsu *et al.*, 2003).

In some countries, most cities and towns have an animal welfare organisation such as the Society for the Prevention of Cruelty to Animals (SPCA) which attempt, by education and action, to reduce the problem of unwanted puppies being produced and to find homes for the dogs surrendered to their shelters. In addition, local government may attempt to control the population of dogs according to national and local legislation.

Local government and veterinary department interest in unwanted dogs is generally based on public expectations and public health. Free-ranging dogs are perceived as a health problem, especially in countries where rabies is endemic, but they may also act as scavengers, cleaning up after humans, removing human faeces and other waste materials. The widespread acceptance of organisations concerned with the welfare of unwanted dogs and the presence of pounds and shelters indicates that people in many societies accept that there is a problem with unwanted, stray and surplus dogs, and the public is willing to pay local government and to fund charities to deal with the problem.

Local government officers enforce both national laws and local government by-laws. These laws are directed towards dog control, dog registration, reduction in public nuisance (fouling, barking, stock chasing) and dog aggression. They often have staff dedicated to catching and impounding dogs until their owner reclaims them or they are re-homed or killed. These organisations are funded for the public good and while they may enforce animal welfare legislation their primary function is to reduce problems that dogs create in society. That dogs are a public nuisance is obvious in that almost every small town in the developed world has a dog pound where stray, unwanted and problem dogs are held until their fate is decided. Staff working in many dog control programmes are often poorly trained, under equipped and their knowledge of the theory of population control, dog capture and handling and euthanasia may be limited.

The simplest dog control programmes are ones in which dogs are killed, or captured and killed. Because dogs are highly fecund, killing programmes have to be well resourced and ongoing so as to kill enough bitches to reduce pup production significantly. If sufficient food is available dogs will migrate to the food supply, and a population of dogs will be maintained, unless a breeding control programme to promote a stable non-breeding population is initiated.

Dog control programmes vary considerably from country to country, depending on the prevailing social attitude towards dogs. In countries where dogs are considered to be important companions of humans, dog control is usually aimed at the elimination of free-ranging dogs, registration, de-sexing and responsible ownership. In countries where dogs are regarded as unclean, then the dog population will be maintained at a level acceptable to the society and if it gets too large then dogs will be killed. In some countries killing dogs is unacceptable and thus unwanted dogs are allowed to multiply and live or die. In the latter societies, de-sexing and habitat change may be the only acceptable methods of population control.

If dogs are eaten then this puts another pressure on the dog population and may be a force in control. In a country in the Middle East, in the 1970s, the author observed a particularly efficient short-term dog control programme when a road construction company from an Asian country purchased dogs from the local community for food. The different attitudes towards free-ranging dogs were well illustrated in a conference on urban animal management held in Amsterdam in 2000. In Chile humans may consider the stray animals in the streets as partners and they are relatively untouchable by the authorities (Tallo, 2000). In Belgrade, the municipality kills free-ranging dogs while the shelters catch dogs to protect them and then keep them alive in overcrowded conditions (Butcher, 2000). In India and in Taiwan, religious beliefs preclude euthanasia as a means of control (Butcher, 2000; Hsu *et al.*, 2003).

The welfare of surplus or unwanted dogs is a major issue with controversy surrounding the methods used to control dog populations and the management of surplus and unwanted animals. The foundation of any dog control programme is acceptance by the community that it has to be done. Control is usually accepted if the community regards rabies or other dog related diseases or behaviour as significant or if dogs are considered to be of sufficient nuisance value to require control. The control programme can then focus on the tools used to control dog population, namely education, dog registration and identification, restriction of movement, habitat modification, de-sexing and contraception, killing, and capture and euthanasia (Table 2).

The reasons for dog control have to be specified as this determines the desired outcome of the programme, for example a rabies control programme may differ from a welfare based programme. Rabies control might concentrate on vaccination and destruction of all non-vaccinated dogs and excess puppy production can be taken care of by natural attrition and destruction of non-vaccinated animals. However, a welfare-based programme could not accept natural attrition that is starvation and disease, as a method of control but would concentrate on reducing the production of unwanted puppies by other means. In many countries forcing owners to take their dogs off the street is an important part of dog control legislation and it is the basis of dog control in Europe and North America. This is very useful in dog control and will have a significant effect in towns and cities worldwide but may not be pertinent to village and rural communities in many parts of the world.

Table 2. The key issues in dog control depend on status of the dog.

Owned Restricted	Owned Free-ranging	Not-owned Free-ranging or Wild
Legislation enforced	Legislation enforced	Legislation enforced
Education	Education	
Registration	Registration	Identification
Habitat Control	Habitat Control	
De-sexing	De-sexing	De-sexing
Contraception	Contraception	
Capture, Reclaim	Capture, Reclaim	
Capture, Euthanase	Capture and Euthanasia	Kill

6.1 Legislation

The development of useful and enforceable laws is fundamental to effective dog control. Enforced registration with penalties for unregistered dogs (Hsu *et al.*, 2003) and routine capture and holding of unregistered free-ranging dogs is important for success.

6.2 Education

Agreement between dog owners and the general public that something needs to be done is essential for dog population and behaviour control programmes. If the majority of dog owners are willing to have their dog identified or, even better, keep it at home, then dog control can begin in earnest. If dog owners want their dog to roam freely then having it identified by a collar or tattoo allows dog control staff to capture non-owned free-ranging dogs and kill them, or hold identified dogs for reclaiming. Registration of ownership may or may not be applicable, depending on the community and its use for, and attitude towards, dogs. It may not be possible to register community owned dogs. However in more and more communities, dog registration fees are used to finance dog control staff and facilities.

6.3 Habitat Control

If food is available for free-ranging and wild dogs then a population will become established to utilise what is available. The presence of garbage including food waste and human faeces on the streets of many cities, towns and villages worldwide provides the nutrition required to sustain free-ranging dogs and many are let loose each day to scavenge for their food in addition to being fed by their owners. Un-owned free-ranging dogs depend on street garbage and dumps for their nutrition. The removal of garbage reduces the food available and this in itself reduces the number of stray dogs an environment can sustain. The removal of street garbage and improved sanitation has significant human health benefits and is a primary responsibility of local government and society. In some countries such as Thailand it is considered appropriate to leave food out for free-ranging animals and then these dog populations are deliberately maintained.

Many free-ranging and wild dogs live on garbage dumps which can be fenced off and/or used as a focus for population control activity either by de-sexing or killing. Fencing off major sources of food will starve a population to death so it is best if fencing is combined with killing, or capture and killing. If fencing is impossible then de-sexing the resident dogs at a dump will produce a defined population in which no more pups are being born. If de-sexed dogs are easily identified, for example by removing a piece of an ear, then new immigrants can be caught and de-sexed, or killed, depending on the desired outcome of the control programme. If more than 90% of a stray dog population is de-sexed then the population may eventually collapse but this is unlikely to happen if dogs can migrate towards a food source from other areas.

Habitat modification is a good tool to use in a dog population control programme. However, some communities use dogs as garbage removal agents and they will want the dogs to do so until there is an alternative. Garbage removal is a costly business and needs to be thorough to reduce the food available for dogs. In many cities and towns in poorer countries it is not carried out very effectively and food is easily available for free-ranging dogs. Removing available food or making access impossible is important if dog control is to be achieved. It may also have huge benefits to local residents with regard to fly and mosquito control and reduction in human disease.

6.4   Killing

The bitch can up to 12 pups per year if resources are available. If a dog population control programme is based on the killing of free-ranging dogs then 50 to 80% of the dog population have to be killed each year for it to be a successful strategy (Kuwert et al., 1985). This is an expensive strategy to use to control rabies but is one tool used in the control of free-ranging dog populations in the wealthy countries of the world where the dogs are captured, held for a period and then killed.

In many Asian countries or communities, people are Buddist or Hindu and killing any animal, including dogs even to control rabies, is seriously opposed (Panichabhongee, 2001) and free-ranging dogs may be fed. In these countries alternative methods are required to control dog populations. In countries where Islam is the predominant religion it may be difficult to get people to handle dogs as to people of that faith dogs are unclean. This makes the killing of dogs by capture and euthanasia difficult. Shooting is dangerous in urban areas and poisoning may be the only alternative. However, the dead and dying dogs have to be picked up and destroyed or buried. In addition, killing free-ranging dogs is often unpopular as many owned dogs are also killed. If it is to be carried out, then human safety and dog welfare must be considered. Wild dogs are usually killed either by poisoning or shooting (see 6.6 Control of Wild Dogs).

In countries where dog registration either does not exist or is not enforced then using killing as the only method to control dog numbers is expensive, difficult and unlikely to succeed. Indeed killing dogs in one area will result in other dogs migrating to that area if resources are available. This movement of dogs is contraindicated if rabies control is one of the reasons why dog population control is initiated, as it will draw dogs into the community and results in social unrest amongst the dogs and may bring disease into the area. In Guayaquil, Equador, about one quarter of the dogs were killed in 1981 and 1982 but it had no long-term effect on the dog population, or the incidence of rabies (Meslin *et al.*, 2000).

The personnel involved in dog control programmes are generally directed to reduce dog numbers in a community. This can be achieved in the short term by the killing or capture and euthanasia of free-ranging dogs. In the long term this will not succeed due to constant immigration unless it is accompanied by habitat modification, education, de-sexing, registration and eventually a change in attitude to how dogs are to be managed.

Capture techniques are important. It may be possible to capture tame dogs with a little bait and a simple loop to put around their neck. More suspicious animals may be captured using a loop on the end of a handle or a net. Wilder animals may need to be trapped using a cage trap. The trap should be baited for a few days to get the animal accustomed to it before it is sprung. Sometimes dogs have to be immobilized using a blow gun or dart gun (Leney & Remfry, 2000).

Dogs should be transported to the pound in an enclosed vehicle and the pound should pen dogs individually to reduce inter-dog aggression. Many pounds are built of concrete with poor conditions for dog comfort. Dogs should be able to sleep on a bed off the ground, to reduce cold and draughts, and the cages should be easily cleaned. Bitches with pups should be held apart from other dogs to prevent aggression and cannibalism, and injured or diseased dogs should be separate and killed as soon as possible to reduce unnecessary suffering.

The word euthanasia is derived from the Greek *eu* meaning good and *thanatos* meaning death signifying 'good death'. It is defined as using a method of killing that causes rapid loss of consciousness with minimal pain or distress and thereafter death. Many methods of killing dogs do not cause rapid loss of consciousness and cause pain or distress. If dogs are to be killed than it is important that humane methods of killing are used. Euthanising agents cause death by one of three methods; hypoxia, direct depression of neurons necessary for life and physical disruption of brain activity, and destruction of neurons necessary for life (Beaver *et al.*, 2001). However, for death to be pain and distress free, loss of consciousness must precede loss of motor activity and therefore agents that cause muscle paralysis without loss of consciousness are not acceptable as sole agents of euthanasia. In many countries the methods used to kill dogs are anything but humane and dogs may be drowned, beaten or starved to death, electrocuted, or poisoned (Butcher, 2000). A wide range of techniques can be used to kill dogs humanely and these are discussed below.

6.4.1   Injectable agents

Barbiturates, which are frequently used to euthanase dogs, are anaesthetic agents. They depress the nervous system starting at the cerebral cortex causing loss of consciousness, anaesthesia and with over-dosage, apnoea and cardiac arrest. A common and a humane way to kill dogs that are used to being handled by humans is to inject the dog intravenously with barbiturates (usually pentobarbitone or pentobarbital sodium). The dog is restrained manually, the drug injected intravenously and the dog rapidly becomes unconscious and dies. There may be some vocalisation during the process. In most countries barbiturates can only be used by veterinarians. The standard dosage rate of pentobarbitone is 150 mg/kg when given intravenously. Some veterinarians use a sedative before giving the barbiturate.

Barbiturates can also be injected intraperitoneally when it is difficult to inject them intravenously. The time to unconsciousness is greater following this route of administration and the dog may become distressed when it starts to become unconscious. Barbiturates may also be given orally; either squirted into the mouth or fed in a bait or meal. When given orally it may take half an hour or longer for the dog to become unconscious and then a further dose of barbiturate can be given intravenously to kill the animal. Pentobarbitone sodium is bitter tasting.

Thiopentone may be used instead of pentobarbitone sodium. It can only be given intravenously and acts more rapidly than pentobarbitone. It is often used in human surgery so may be available when pentobarbitone is not but it is usually more expensive.

If the dog is difficult to handle, because it is afraid of humans or is dangerous, it may have to be caught with a dog pole and sedated using an intramuscular injection given by hand or using a pole syringe. Then when it is safe to handle, the dog may be killed using an intravenous injection of a barbiturate. Xylazine may be used as an intramuscular sedate at a dose of 1-2 mg/kg (Reilly, 1993).

T61 is a mixture of three different chemicals that act as a general anaesthetic, a curariform drug and a local anaesthetic. It has been used as a substitute for barbiturate, but if administered at the wrong speed it can cause muscle paralysis in the still conscious animal. For this reason it is not available anymore in the USA but is available elsewhere.

A dog may be anaesthetised with other injectable drugs such as a ketamine/xylazine mixture or by inhalation of gaseous anaesthetics (see below) and can then be killed with a variety of chemicals including magnesium sulphate ($MgSO_4$) delivered intravenously as a concentrated salt solution. $MgSO_4$ may cause the bowels to empty and so using it for euthanasia can be a dirty procedure. Potassium salts such as KCl, KNO_3, KSO_4 can also be given intravenously to kill anaesthetised dogs (Leney & Remfry, 2000).

It is important after using intravenous drugs, to ensure that the dog is dead. There must be no respiration, no heart beat, no corneal reflex and the eyes should have glazed over.

6.4.2 Inhalation agents

Gaseous anaesthetics such as halothane (Reilly, 1993) can be used to anaesthetise smaller dogs which then may be killed by shutting off the oxygen supply or using any of the drugs listed above. Animals may struggle during anaesthesia induction and Beaver et al. (2001) recommended that, in order of preference, halothane, enflurane, isoflurane and sevoflurance are acceptable for use in dogs that weigh less than 7kg.

Carbon monoxide (CO) is a colourless and odourless gas that is not explosive below 10% concentrations. It is used in many shelters to kill dogs. It is highly poisonous and dangerous to humans. Pure CO can be purchased as a compressed gas in cylinders and used through a special unit. It is also present in the exhaust fumes of cars and these can be cleared and cooled through a set of filters but there may be problems with concentrations and other gases so that only CO from gas cylinders should be used (Beaver et al., 2001). When CO was used to kill dogs there was a 20 to 25 second period of abnormal cortical function and dogs vocalised during this time and became agitated (Chalifoux & Dallaire, 1983). Humans often use CO for suicide. CO poisoning in humans is associated with dizziness, nausea and headaches so it probably causes some distress in dogs before they lose consciousness.

Carbon dioxide (CO_2) at 30 to 40% concentrations has been used to induce anaesthesia in dogs, within one to two minutes, usually without struggling, retching or vomiting. Beaver *et al.* (2001) recommended compressed CO_2 in cylinders for use as a euthanising agent for dogs.

Nitrogen (N_2) at a concentration of 98.5% caused dogs to become unconscious after 76 seconds and their electroencephalograms became isoelectric after 80 seconds. Following loss of consciousness the dogs vocalised, gasped and convulsed and some had muscle tremors. Herin and colleagues (1978) concluded that it was a humane method of euthanasia but Beaver *et al.* (2001) did not recommend it for dogs as it may cause hypoxia before loss of consciousness.

6.4.3   Firearms

A restrained dog can easily be killed using a revolver, pistol, shotgun or small bore (0.22 calibre) rifle or a captive bolt pistol. The captive bolt pistol should be held against the dogs head between the eyes and ears. It may only stun the dog which then has to be bled out. Captive bolt pistols may be difficult to use as the dog may not hold steady and the heads of small dogs may make it difficult to use effectively.

A pistol should be held against the head but other firearms should be held about 5 cm away from the head. There is a danger with using revolvers and rifles of the bullet ricocheting against the floor or wall if shooting is carried out inside or in a concrete enclosure. A small gauge shotgun (410) is safer to use and very effective if held 10 to 20 cm away from the head.

6.4.4   Clubbing

The idea of clubbing a dog to death is unacceptable to many people but a well placed blow from a weighted pipe to the head can incapacitate the dog making it unconscious and a second blow result in death. A confident, well trained, capable person can dispatch dogs humanely using this method. However poor delivery of the initial blow can result in a very inhumane death and this method is not recommended.

6.4.5   Electrocution

Electrocution using alternating current has been used to kill dogs. It causes death by cardiac fibrillation and animals may not lose consciousness for 10 to 30 seconds after the onset of fibrillation, therefore, it is better if animals are unconscious before electrocution (Beaver *et al.*, 2001). Electrical stunning can also be used but the electrical current has to be directed through the brain to induce rapid loss of consciousness (Beaver *et al.*, 2001). If the current passes only between fore and hind limbs, or neck and feet, it causes cardiac fibrillation without a sudden loss of consciousness (Roberts, 1954). Boxes designed to pass an electric current thorough the brain to render it unconscious and cause cardiac arrest used to be manufactured. Their use was considered humane by Leney and Remfry (2000) but they are no longer manufactured.

There are many other methods of killing dogs that are not recommended because they cause serious pain and distress or do not cause the animal to lose consciousness quickly. Drowning has been commonly used in the past to kill dogs especially pups. Chloroform chambers were also used to kill pups but their use was discontinued for human health reasons as chloroform is toxic. Curare drugs should not be used as they cause paralysis of the respiratory muscles and this is known in humans to be very distressing.

Beaver *et al.* (2001) recommended barbiturates, inhalant anaesthetics, CO_2, CO and potassium chloride in conjunction with general anaesthetic as acceptable means of euthanasia in dogs. They suggested that N_2, Ar, penetrating captive bolt and electrocution were conditionally acceptable. A major problem with the barbiturates and inhalant anaesthetics is that they can be used by veterinarians only and they are generally expensive to purchase and to deliver. Thus in pounds and shelters where funds are scarce and equipment non-existent they cannot be used. Reilly (1993), in reviewing euthanasia of animals used in research, recommended pentobarbitone either intravenously or intraperitoneally and accepted with reservations halothane, CO, CO_2 and captive bolt. Ether, chloroform and hydrogen cyanide were unacceptable to Reilly (1993) because they are a danger to use. Decompression and electrocution were unacceptable as being possibly inhumane, aesthetically unpleasant, and require specialist equipment.

6.5   Control of reproduction

Where there are difficulties with using killing as a population control method and when a stable, controlled population of free-ranging dogs is acceptable, then de-sexing or contraception may be a suitable method to reduce reproductive success and establish a defined and controlled population of dogs. Free-ranging dogs are useful to people living in many countries in that they act as sentinels outside houses and dispose of garbage. They may not be owned but they are valued, and controlled populations may be the optimum outcome. Contraceptive methods can be applied to the female or male or both depending on the programme developed and what is socially acceptable. However many people object to their dog being de-sexed and men especially object to having their male dog castrated.

6.5.1   Female

The simplest way to prevent a bitch having pups is to keep her indoors away from male dogs. Specially designed pants or intravaginal plugs may be used on individual animals, but these are subject to failure. In addition reproduction may be controlled surgically by ovariohysterectomy (the removal of the ovaries and uterus), hysterectomy or ovariectomy alone, the use of contraceptive pills or injections, and immunocontraception (Lohachit & Tanticharoenyos, 1991; Fayrer-Hosken et al., 2000).

Surgical de-sexing of females is the best method of population control. It is permanent and while the animal maintains her status and position in the community of dogs she is now sterile. Indeed her health may improve and she may attain a higher position in the pack and will live longer than if she continued to breed. Because a general anaesthetic is needed and the surgical procedure is complex, de-sexing of the bitch has to be carried out by a veterinarian. The surgical risk is small if the veterinarian is trained correctly. In many developing countries veterinarians focus on livestock production and health, and may not concentrate on dog health. Minor modification to the curriculum in most veterinary schools may make their graduates capable of carrying out an ovariohysterectomy and if they have the means available then they can carry out this procedure anywhere. It is also important to stress the population dynamics of dogs as many veterinarians may believe that capture and killing is a more effective way of population control. Ovariectomy may also be carried out rather than ovariohysterectomy and the former is a simpler surgery with fewer post-operative problems. The incision is shorter, there is less trauma in the abdomen, and the broad ligaments are not torn and the surgery is much shorter. Pyometra is apparently not a problem (Janssens & Janssens, 1991; Okkens et al., 1997; Veenis, 2004). Ovariectomy may be the technique of choice in developing countries wanting to maximise de-sexing at minimal cost.   Hysterectomy is not recommended as it results in the bitch continuing to cycle with all the attendant social problems.

With good anaesthesia, surgical technique and post-surgical management mortality following ovariohysterectomy can be almost zero. The major cause of mortality is poor equipment, technique and training. There is a link between ovariohysterectomy and urinary incontinence (Gregory, 1994) but it also has a positive effect in reducing the likelihood of mammary tumours developing. There is a greater incidence of obesity in spayed bitches (Anderson, 1973; Edney & Smith, 1986) but this is unlikely to be a problem with free-ranging animals. Increasing the longevity of bitches is important as it makes for long-term stabilization of the population. Spaying may make bitches more aggressive and this may make them more successful foragers and also increase their longevity. The pain after surgery can be reduced by systemic analgesia (see Chapter 6).

There are several hormonal treatments available to reduce conception in bitches. These are useful in owned animals and for short term population control but longer term treatment may cause unwanted side-effects (Evans & Sutton, 1989). The development of longer lasting hormonal contraception techniques in the bitch, such as the use of long lasting gonadotrophin releasing hormones, show promise (Trigg *et al.*, 2001).

In Thailand reproduction control has become an integral part of the rabies control programme and in 1996 660,000 bitches received contraceptive hormonal injections and 55,000 were spayed. It is not known what the effect of this will be on the population in the long term (Meslin *et al.*, 2000)

An ideal contraceptive would be a long-acting drug, effective for 3 or 4 years, that could be given orally in a bait to free-ranging dogs. This would be particularly suited to countries such as Thailand where killing is not well accepted. Oral administration would be much better than an injectable format as a large percentage of bitches would have to be contracepted to stabilize the population and prevent pups being born. Free-ranging dogs will easily take food baits but catching and injecting them is much more time consuming and less effective.

There are two types of immunocontraceptive vaccine; one generates antibodies against the zona pellucida (ZP), the other antibodies against luteinising hormone-releasing hormone (LHRH) which controls the release of male and female hormones like testosterone, progesterone and oestrogen. Both are usually temporary in action and need to be given every 6 to 12 months. In ZP immunocontraception, the bitch is vaccinated with ZP glycoproteins leading to immunocontraception (Olson & Johnston, 1993). The method works by stimulating the production of anti-ZP antibodies which then block fertilization. This form of contraception does not stop cycling and so bitches will come into heat but not conceive. The LHRH vaccine can be given to male or female dogs rendering them infertile for 6 months. It is also possible to cause abortion in pregnant bitches using a number of hormones.

6.5.2   Male

There are four principal methods of reproduction control in the male dog; surgical castration, vasectomy, chemical sterilization by injecting a chemical intra-testicularly or into the epididymis (Olson & Johnston, 1993), and castration by burdizzo or rubber ring (Lohachit & Tanticharoenyos 1991).

Surgical castration is a guaranteed method of reducing fertility in male animals. Surgical castration of dogs should be carried out under general anaesthesia which limits the procedure to veterinary surgeons. The pain experienced after recovery is likely to be significant and more veterinarians are using systemic analgesics to reduce this pain. Lohachit & Tanticharoenyos (1997) suggested that a modified human vasectomy clamp could be used as a fast and easy method for mass sterilization of dogs. General anaesthesia would still be required but if it is a more rapid procedure then it might be suited to public programmes. Vasectomy may be an effective means of contraception but the dog will continue to mate and may be involved in aggression with other dogs which may result in injury. That the testicles remain might satisfy those owners, usually male, who object to having their dog castrated.

In chemical castration any one of a number of chemicals is injected into the testicles, ductus deferens or epididymis resulting in azoospermia (Olson & Johnston, 1993). The pain caused by chemical castration of dogs nor its efficiency has been documented but in cattle chemical castration was painful and also not very effective and it is probably not suited for use in dogs. The pain caused by burdizzo and ring castration has not been defined in dogs, nor has their efficacy been demonstrated. Rubber rings should not be used to castrate dogs as it takes several weeks for the scrotum to detach. The use of the burdizzo is also questionable. To establish a stable population long living dogs are required and castration may facilitate this better than vasectomy.

In developed countries a large percentage of dogs are de-sexed. In Ontario 66% of dogs were de-sexed and 20% of the remainder were kept for breeding whilst 14% had reproduced (Leslie *et al.*, 1994). However in Sweden, less than 2% of dogs were de-sexed (Sallander *et al.*, 2001) and the presence of unwanted pups was not a greater problem than elsewhere. De-sexing is useful when it is difficult to restrict dogs to an apartment or household and when many dogs are free-ranging. All de-sexed animals should be marked with either a tattoo in the ear or an ear notch if returned to a free-ranging state so that they can be easily identified. De-sexing male dogs is not as effective as de-sexing females and if resources are limited then de-sexing programmes should concentrate on the females.

6.6   Control of wild dogs

Wild dogs are animals that have left human society and generally live removed from human settlement. They may prey on livestock and are frequently targeted by farmers. In Australia wild dog control is carried out by shooting and poisoning. Dogs can also be used to kill wild dogs.

When guns are used to kill free-ranging dogs the type of weapon used depends on the location of the dogs. All firearms are dangerous and high calibre rifles should not be used in built up areas. Good marksmen can use a

12 gauge shotgun if the dogs are to be shot within 20 to 25 meters and 0.22 calibre rifles are effective if the bullet is placed correctly. Shooting dogs is similar to shooting any animal. A clear field of fire is required and the animal should be shot in the head, neck or heart area. However, a gunshot to the neck or heart may not immediately render an animal unconscious (Beaver *et al.*, 2001). Marksmen should be skilled and a rifle of sufficient calibre should be used. Although 0.22 calibre rifles are probably sufficient, larger calibre rifles will always have a greater impact and are more suited for dog shooting in rural areas.

Meat baits containing 1080 (Sodium monofluoroacetate) are used for poisoning wild dogs and dingoes in Australia (Fleming, 1996). The percentage of dogs or dingoes killed, varies from 10% (Bird, 1994), 22% (McIlroy *et al.*, 1986) to 69% (Best *et al.*, 1974) and above 70% (Fleming, 1996). Aerial baiting with 1080 poisoning is also a very effective method of killing dingoes and wild dogs (Thomson, 1986; Fleming *et al.*, 1996). Dogs are highly susceptible to poisoning with 1080 with an LD50 values of 0.07 mg/kg bodyweight which is much lower than values for cats (0.2-0.3) (Rammell & Fleming, 1978). In New Zealand dogs are occasionally poisoned accidentally by 1080 when scavenging on carcasses of possums (*Trichosurus vulpecula*) poisoned by it or even by eating or licking cereal baits placed for possums (Meenken & Booth, 1997). In dogs 1080 causes severe and enduring clinical signs and it takes some time for the animal to become comatose.

Cyanide causes rapid death in dogs and is a very useful poison for use in dog control programmes. It is highly toxic to humans and needs to be used with care. Strychnine is also used to kill dogs (Best *et al.*, 1974) but has little to recommend it. A small amount is sufficient to kill a dog but the dog poisoned with strychnine goes into convulsions and endures a painful death. There is little evidence on the effects of other poisons on dogs except veterinary experience of warfarin and phosphate In New Zealand, trials are being carried out to combine poisons with analgesics and sedatives to reduce the pain and distress experienced by brush tail possums following poisoning.

7.   CONCLUSIONS

There are many unwanted free-ranging dogs and these may be a public nuisance and a source of zoonotic diseases, especially rabies. These dogs are often underfed, diseased, and victim to road traffic accidents and abuse and they produce many pups that die while very young. The desire to control rabies and other diseases may encourage countries to control these dogs. The mechanisms to do so (education, registration and identification, habitat control, de-sexing and contraception, killing, capture and euthanasia) are well known but the political will and financial capacity to do so are often lacking as is the public support for such programmes. The ongoing existence of public dog pounds in most developed countries illustrate the difficulties of effective control of free-ranging dogs and the apparent disregard of a percentage of dog owners for the public nuisance caused by their dogs. If there is an ongoing problem in wealthy countries then rapid changes in much poorer countries cannot be expected. The development of a cheap, easy to administer and long term contraceptive might reduce the production of unwanted puppies from treated bitches but to be effective about 90% of bitches would probably have to be treated. The quality of life of free-ranging dogs, their capture and euthanasia are the major welfare issues for dogs worldwide. Support by animal welfare organisations in developed countries for veterinarians and welfare organisations in developing countries to help control these dogs will have a significant effect on dog welfare worldwide.

Chapter 3

BREEDS AND BREEDING

Abstract: There are several hundred breeds of dogs and they are usually categorised according to function. Many of these dogs are now companion animals and do not engage in what they were initially bred for. Pedigree dogs used for breeding are registered by a kennel club. Most dog breeders have one or two breeding animals and they try to produce puppies that meet the breed standard for physical and behavioural characteristics. Some of the breed-standard physical characteristics are extreme and may cause welfare problems. Inbreeding and line-breeding have tended to increase the incidence of hereditary diseases. Emphasis on physical characteristics may have led to breeders paying insufficient attention to behaviour. The physical and behavioural characteristics of some breeds make them unsuited as companion animals. There is now a more definite effort being made to reduce the incidence of hereditary diseases in some breeds but the costs of developing and using diagnostic tests may make it difficult for breeders to utilise them effectively. Genetic counselling for breeders is a major development in veterinary science and will hopefully reduce hereditary disease while maintaining breed characteristics. Minor changes in the physical breed standard of some breeds could have significant effects on the welfare of pedigree dogs.

1. INTRODUCTION

There are more than 400 breeds of dog, each developed to carry out a particular activity. Breeds are categorised by national kennel clubs according to original function or type, but they can also be categorised genetically. Breeds can be divided into two major groups, ancient breeds and modern European breeds, using genetic variation (Parker *et al.*, 2004). The latter breeds can be subdivided into three groups, mastiff-type dogs, collies and

Belgian sheepdogs, and hunting dogs. The ancient breeds, which include the Shar-pei, Shiba-inu, Chow Chow, Akita and Basenji also include two gazehounds (Saluki and Afghan Hounds) and sled dogs (Siberian Husky and Alaskan Malamute), have been around for thousands of years. Many breeds, however, were developed to meet particular needs, especially hunting needs. In Europe, there were two major periods of dog breed production, the Middle Ages and the 19th Century. In the Middle Ages, hunting was a symbol of power and the aristocracy produced different types of dogs for different game and types of hunting. They bred Deerhounds, Wolfhounds, Boarhounds (Clutton-Brock, 1995) and Beagles/Harriers. Later, Foxhounds were produced as foxes became an important game animal and hunting from horseback became common. Gun dogs were developed as shooting gamebirds became popular and took over from netting birds. In addition, there was a host of small dogs for hunting above and below ground (terriers, dachshunds), sheep and cattle dogs of several types, guard dogs and a few toy breeds.

In the 19th Century breed development became very popular and breeds with documented pedigrees were developed (Ott, 1996). Dog shows became popular and the first dog show was held in England in 1859 in Newcastle upon Tyne. The Kennel Club in England was established in 1873 while the American Kennel Club was established in 1884 (Anonymous, 1998a). There was a surge in the production of different breeds of dogs which had a local distribution were identified as breeds. Their physical and behavioural characteristics were listed and breed books opened. Breeds were developed through such intensive inbreeding and line-breeding that many breeds can now be identified by their genotype (Parker et al., 2004). In the past, the aristocracy held personal records of their dogs' pedigrees but in the 19th Century these pedigree books became public. In the UK and Ireland, more than 20 breeds of terrier were identified, the majority of which did the same type of work so the different breeds were classified primarily on physical characteristics. Some dog breeds were re-established. For instance, the Irish Wolfhound was extinct in Ireland but was reproduced by Captain George Graham, from a mixture of large breeds, despite there being no wolves to kill. He began a breeding programme in 1862 and a breed standard for the Irish Wolfhound was produced in 1885 (Anonymous, 1998a).

The origin of many breeds of dogs is unknown, but four elements (founder animals, isolation, inbreeding, selection) are required to develop a new breed genetically. There is an initial group of founder animals, usually related in ancestry, type and function. This group is then isolated genetically from other dogs and bred together to form the breed. The founder group is usually a small group of dogs and inbreeding is needed to stabilise the physical and behavioural characteristics of the new breed. Thereafter, selective breeding for specific characteristics allows the breed characteristics to become established genetically.

The breeding of pedigree dogs is under the control of kennel clubs which hold the description of each breed and records the pedigree of each registered dog. When dog showing began dogs were expected to carry out specific activities to a certain standard before they were identified as champions. This became difficult as dog breeding became more widespread and showing dogs more democratic. Soon, champions were identified according to their physical characteristics and behaviour in the show ring. This allowed refinement of the physical characteristics of each breed and permitted breeders to move away from the original function of the breed. This was to be expected as more and more dog sports (dog fighting, bull baiting, badger baiting, hare coursing) became illegal or difficult to undertake. Showing dogs, and breeding for this pursuit had a major impact on dog ownership. Dog breeding became widespread and many people were involved. Breeding pedigree dogs became a sport and a minor industry. To underpin the costs of showing, many people breed from their best animals to generate a little income. Each breed and its varieties has a set breed standard and breeders and showers try to produce a dog as close to the ideal as possible. A champion, valuable as a stud dog, could generate stud fees and the pups from champion bitches were sought after.

Field trials for most breeds became impossible and only a few types of dogs, principally gun dogs and sheep dogs, are now trialled extensively. This has resulted in different strains of dogs being bred for showing (pedigree), for field trialling (pedigree or non pedigree) and for work or companionship (often non-pedigree but purebred). In the UK, this happened to the Border Collie in the last few decades. To become a full champion, Border Collie show dogs had to pass a field test but few collies have taken the test and fewer have passed it and so the 'show' Border Collie will lose its ability to work sheep (Willis, 1995). This may be a good thing as good working Border Collies are extremely focussed on particular types of behaviour and are not suited to be companion animals.

New breeds continue to be recognised by kennel clubs. Since 1990, the American Kennel Club has registered 20 new breeds out of a total of 150 registered breeds. Breeds are also being developed. In New Zealand, two types of dog, the Huntaway, used to drive sheep, and the Heading Dog, which directs sheep, are types of dogs with a wide range of physical characteristics. The New Zealand Kennel Club has suggested that the former become registered as a breed. This suggestion dismayed many because Huntaway exists to do a specific job and has a wide variety of physical characteristics. It is not suited to be a companion animal, having a predisposition to bark.

Controlled production of pups bred according to breed standards is necessary if dogs with known physical characteristics and behaviour are to be available for purchase. Retention of a range of breeds to suit the requirement of different people for work, sport and companionship is a major outcome of the work undertaken by kennel clubs internationally. Many working dogs are registered with kennel clubs but are never shown. Dogs bred for specific purposes such as guide dogs need to have their pedigrees recorded and controlled and kennel clubs remain the best way of maintaining these records.

There are many aspects of breeding dogs for show competitions and the process of showing itself which impact directly or indirectly on the welfare of these animals. These include:

1. Dogs may be made into champions and bred from before they have matured physically and behaviourally.
2. The use of artificial insemination may allow animals that have serious physical or behavioural problems to be bred.
3. Emphasis on physical characteristics and behaviour in the show ring does not necessarily produce dogs suited for companionship and most pedigree dogs are now owned as pets.
4. The physical aspects of some breed standards may predispose animals to injury or disease.
5. Inbreeding or line-breeding may increase the incidence of hereditary diseases, particularly if the disease status of many important stud dogs and brood bitches is unknown.
6. Selection for specific physical characteristics may be accompanied by unwelcome behaviours or an inability to perceive or communicate effectively.
7. Limited gene pools may not allow selection away from hereditary problems and may increase physical, behavioural and disease problems in a breed.
8. Breeding for types that will be modified surgically, so that tail and ear features are ignored.

However, pedigree dog breeders and showers are usually determined to improve and maintain the best characteristics of the breed they are involved with. This enthusiasm, coupled with the rapidly developing field of diagnostic genetics in veterinary science, genetic counselling, and the increasing ease of preserving dog semen and shipping it worldwide, will make it easier for kennel and breed clubs to develop programmes to improve their breeds and guarantee their futures. Dog breeders are frequently criticised but most wish only to improve their breed and there are modern techniques which can help them achieve their goals and thus improve the welfare of their dogs.

In this chapter, welfare aspects of breeding programmes, physical and behavioural characteristics, and inherited diseases will be discussed.

2.   BREEDING

The promiscuity and fecundity of dogs is both a strength and a weakness. Dogs reach puberty by 6–9 months of age and this allows them to be bred before they are physically or behaviourally mature. The ability of many bitches to breed twice-yearly makes it possible to produce a large number of puppies from one bitch over a few years, and because litters can be large a few animals, particularly stud males, can dominate a breed. A dog can become a champion before 2 years of age, thus not allowing the true behavioural characteristics of an animal to be seen before breeding, as many dogs are not socially mature until 18–36 months of age (Overall, 1997).

In most countries, there are breeds with low numbers of animals. In this situation, individual animals may be in-bred continuously as there may be no opportunity to breed to an unrelated dog. Many less popular breeds get into a cycle of breeding closely-related animals because the demand for the breed is low and the number of dogs in the register is low; this is a problem in all countries. In the USA, 150 dog breeds are listed with the American Kennel club and in 2002, 958,503 pedigree animals, including pups, imported animals and late registrations, were registered. The most popular breed, the Labrador Retriever, had 154,616 new registrations but the 50 least popular breeds had less than 515 and the 10 least popular less than 100 new registrations (Table 1). In Australia in 2002, 14 of 188 breeds and varieties had no puppies registered. In isolated countries such as New Zealand, with low populations of some dog breeds, the problem may be greater as it becomes very expensive and is economically not worthwhile to import new bloodlines. In New Zealand in 1995, of the 169 breeds listed with the kennel club 39 had no litters at all (Table 2).

Table 1. The number of new registrations (pups, imported animals, late registrations) by the American Kennel Club in 2002 of the 10 least common breeds.

Breed	Nunber of registrations
German Pincher	76
Plott	72
Foxhound (American)	68
Komondorok	65
Skye Terrier	59
Finnish Spitz	52
Ibizan Hound	50
Foxhound (English)	38
Harrier	23
Otterhound	17

If it is assumed that virtually every animal has some deleterious genes and many dogs may have up to 20 (McGreevy & Nicholas, 1999), then inbreeding and line-breeding will increase the frequency of these genes and novel defects will appear regularly. When breed populations are low, these effects can be disastrous. Recent developments in the preservation of dog semen have made the international movement of semen, and therefore bloodlines, easier and should reduce the problem of inbreeding. However, in reality many of the minor breeds are scarce everywhere. As an example, the Irish Water Spaniel had 116 new registrations out of nearly a million in the USA in 2002, 121 of 245,904 in the UK, and 12 of 66,700 in Australia in 2003. Those small numbers of puppies added to the population contrast with the 150,000, 41,306 and 5,134 Labrador Retrievers registered in those countries, respectively. Some breeds, like the Irish Water Spaniel may have quite limited usefulness. It is not well suited to be an urban companion animal as it is large, often very excitable, hairy, and may have an oily smell. Its role as a retriever of killed and wounded waterfowl has been taken over by the Labrador Retriever, which is much easier to maintain, and its general bird-hunting activities can be carried out by any one of several more popular breeds.

Table 2. The number of litters produced by the 169 breeds registered with the New Zealand Kennel Club in 1995.

Number of litters	Number of breeds
0	39
1	6
2	17
3	7
4	5
5	4
6	6
7	5
8	4
9	6
10	2
Less than or 10	101 (60%)
11-15	13
16-20	9
21-30	13
31-40	14
41-50	4
11–50	53 (31%)
51-100	11
101-200	2
201-300	2
More than 50	15 (9%)

Although artificial breeding can be used to improve a breed through the infusion of new bloodlines from different countries, there are problems with out-breeding in that if the chosen male is not of high quality or is the carrier of a hereditary disease, perhaps as an autosomal recessive trait, it can then spread this problem through a small hitherto healthy population. The choice of sire to be used in a small, often inbred population is extremely important with regard to its morphology, disease status, behaviour and working ability, as it will have a great impact on the isolated population (Swenson, 2001). Moreover, artificial insemination can be used to allow male and female animals, that are incapable physically or behaviourally of breeding naturally, to reproduce. This may result in poor conception rates (Linde-Forsberg, 2001) or the production of faulty offspring. The use of artificial insemination for animals that cannot or will not breed is an important welfare issue as it may increase the percentage of animals of that breed with physical or behavioural problems that render them incapable of breeding. Semen quality may also be poor in some breeds, such as the Irish Wolfhound (Dahlbom *et al.*, 1995), and intrauterine artificial insemination under general anaesthesia may be required; whether this is acceptable or not is debatable (Linde-Forsberg, 2001).

Table 3. New registrations of "fashionable" dogs in Australia (1986–2003) and the United Kingdom (UK) (1994–2003).

	Siberian Husky		Rottweiler		Dalmatian	
Year	Australian	UK	Australian	UK	Australian	UK
1986	160	5,005	NA	1,003	NA	NA
1987	137	6,378	NA	1,222	NA	NA
1988	199	6,851	NA	1,124	NA	NA
1989	301	7,095	NA	1,214	NA	NA
1990	273	8,928	NA	1,007	NA	NA
1991	449	5,965	NA	1,089	NA	NA
1992	564	5,666	NA	999	NA	NA
1993	618	5,935	NA	1,194	NA	MA
1994	950	452	5,314	3,070	1,300	2,794
1995	1,005	576	4,659	3,597	1,472	3,120
1996	1,223	736	4,527	4,148	1,452	3,910
1997	1,182	614	2,830	4,561	1,292	3,786
1998	1,659	796	3,594	4,954	1,538	3,058
1999	1,344	739	2,430	5,306	1,018	2,679
2000	1,371	829	2,288	5,226	1,018	2,752
2001	1,149	1,038	2,111	5,587	896	2,062
2002	1,115	985	1,858	5,802	804	2,071
2003	866	1,491	1,839	6,369	767	2,253

NA = not available

As some breeds decline in popularity others become popular. In Australia, the Siberian Husky and Rottweiler became popular in the 1990s but are now in decline, although their numbers continue to increase in the UK (Table 3). Another breed, the Dalmatian, became popular in the UK and Australia in the 1990s, probably due to the movie '*101 Dalmatians*', but is now also in decline numerically (Table 3). Physical beauty and fashion encourage people to select a particular breed and the fecundity of most breeds allows the demand to be met. Breed societies with re-homing clubs and animal shelters often retrieve dogs when they become unwanted as adults and require re-homing. Breeds become fashionable regardless of their suitability for being pets, but this does not seem to be influenced by the show success of particular breeds (Herzog & Elias, 2004). Some breeds which were once very popular appear to be in rapid decline. For example, the Yorkshire terrier had more than 12,000 puppies registered in the UK in 1994 but only 4,073 in 2003.

The production of pedigree puppies appears to be declining in some countries. In Australia, 95,792 puppies were registered in 1986, 89,922 in 1998, but only 66,710 in 2003. This contrasts with an apparent increase in the overall dog population from 3.1 million in 1998 to 3.6 million in 2003. In New Zealand, registration of pedigree puppies has declined from 11,970 in 1996 to 10,627 in 2002, while numbers of dogs have declined from 610,000 to 594,000 over the same period. In the UK, pedigree dogs make up about

60% of the population (Anonymous, 1998b). There registration of pedigree puppies has remained steady at 246,053 in 1994 and 245,894 in 2003 (UK Kennel Club Data) while numbers of dogs have declined. Thus, in different countries the number pedigree dogs may be decreasing, static or increasing.

In some breeds with low populations, the degree of inbreeding will result in an increase in the number of animals with inherited defects. If litters are produced infrequently and a small percentage of dogs in the population are used for breeding, then the dangers of inbreeding are kept low. The ease of transporting semen from country to country should reduce the problem of inbreeding, even in breeds with low numbers. This activity should be encouraged by national kennel clubs. In addition, in most countries it is possible to take a bitch to be bred in another country, thus in effect increasing the genetic variation within a breed (Swenson, 2001). An alternative strategy for kennel clubs would be to remove breeds from their lists if and when their numbers fall below a certain level. This would result in breeds becoming extinct in some countries.

When the ideal dog of any breed is produced with perfect physical and behavioural characteristics and is disease-free, there may be a desire to clone it. The technology of cloning mammals, and especially dogs, is at an early stage but will become easier within the next decade. The demand to clone much loved pet dogs is also growing. There are specific welfare issues with the cloning process itself in that many newborn cloned animals may not survive for long.

3.   PHYSICAL CHARACTERISTICS

Originally, the physical characteristics of a particular breed were determined by its function. Thus, terriers were small to go to ground while the gazehounds were long-legged to run quickly. Individual breeds of terrier or gazehound were produced for specific local hunting conditions. The large, hairy Borzoi was used to hunt wolves in Russia. The Greyhound was used to course hares and is smaller and more agile. Airedale Terriers were used above ground and are taller than Sealyham Terriers, which were used to hunt badgers and foxes underground. These physical characteristics may have been extreme but the animal had to function, fit its purpose, breed naturally, remain healthy, and be reasonably long-lived. During the initial phase of a breed's development and the production of a breed pedigree, specific characteristics which distinguished that breed were identified and bred for. Thus, many types of terriers were identified as definite breeds and were bred for specific characteristics, although many had roughly the same functions (Table 4) This period of production of a breed for its own sake was considered 'toxic' by Ott (1996). Some have suggested that there are too many breeds (Prole, 1981), and that breeds, such as the terriers listed in Table 4, with similar functions and types should be crossbred to produce fewer breeds with much greater genetic variation.

However, breed differentiation and breeding for particular characteristics are some of the pleasures of owning the dog as a companion animal and has enriched the world we live in. As long as these characteristics were limited by the function of the breed then they remained of benefit to man and animal. However, when showing became distinct from work then characteristics which might be a feature of the original breed type sometimes became exaggerated, and in some have reached a stage where they would be an impediment in the field rather than either a necessary feature or an irrelevant breed characteristic. For example, the excessive hair cover of the Old English Sheepdog, particularly of the face area, is never or rarely seen in types or breeds of working sheepdogs. If dogs are not going to be used for the purpose for which they were initially developed then it might be argued that the physical characteristics required for that activity are no longer needed and therefore the standard could be changed towards a more 'normal' morphology. The size and shape of the skull of baiting breeds, such as the bulldogs, and the location of their eyes and nose could be relaxed to a more dog-like, that is dingo-like, shape and position. However, the opposite has tended to happen and breed characteristics have become exaggerated rather than relaxed.

The breed standard for some breeds may encourage breeders to select for characteristics that result in physical or behavioural problems. The list of these is long and McGreevy and Nicholas (1999) identified a few. For instance, the breed standard for the Pug recommends eyes to be very large bold and prominent (Anonymous, 1998a) and as a result many Pugs are presented to veterinarians with exophthalmoses and exposure keratitis. Other physical characteristics that predispose to disease include shape of the ears and otitis externa in some breeds of spaniels, loose skin on the head in Basset Hounds leading to drooping eyelids, excessive skin folds and skin resulting in dermatitis and eyelid problems in the Chinese Shar-pei, and length of the back and back problems in Dachshunds (Priester, 1976). Some breeds with deep chests may be predisposed to gastric torsions. Breed standards could be adjusted to direct breeders towards physical types with less of a predisposition towards disease. McGreevy and Nicholas (1999) argued that some breed standards are confusing and quoted from the UK Kennel Club standards for Shar-peis. This breed must have loose skin and a frowning expression but the eyelids should in no way be disturbed by the surrounding skin folds and should be free from entropion; McGreevy and Nicholas (1999) argued that the features listed predisposed to entropion.

Table 4. The general functions of British and Irish breeds of terriers.

Breed	Hunting underground	Small	Target species* Medium	Large	Fighting	Stock work
Airedale		√	√	√		
Bedlington	√	√	√			
Border	√	√	√			
Bull					√	
Cairn	√	√	√			
Dandie Dinmont	√	√				
Smooth-haired Fox	√	√	√			
Wire-haired Fox	√	√	√			
Glen of Imaal	√	√	√			
Irish	√	√	√	√		
Kerry Blue	√	√	√	√		√
Lakeland	√	√	√			
Manchester	√	√	√			
Norfolk	√	√	√			
Norwich	√	√	√			
Parson (Jack) Russell	√	√	√			
Scottish	√	√	√			
Sealyham	√	√	√			
Skye	√	√	√			
Soft Coated Wheaten	√	√	√			
Staffordshire bull					√	
Welsh	√	√	√			
West Highland White	√	√	√			

*Small = rats; Medium = fox, badger, otter; Large = deer, large cats, pigs

Some breeds have well-known congenital defects. The Dalmatian is often deaf; 18.4% of a sample of 1,234 Dalmatians were deaf (Wood & Lakhani, 1998) and of these 13% and 5.3% were unilaterally or bilaterally deaf, respectively. This common problem is not mentioned in the description of the breed in the Complete Dog Book published by the American Kennel Club (Anonymous, 1998a). Many dogs that are merle-coloured are deaf and/or have eye problems (Klinckmann *et al.*, 1986). In one study of 1,679 pups arriving at a pet store, Ruble and Hird (1993) identified congenital defects in 253 (15%) of them when 6 to 18 weeks of age. Some had more than one defect. In their classic study of the behaviour of dog breeds, Scott and Fuller (1965) found serious physical defects in every breed studied although good breeding stock was used.

In last three decades of the 20th Century, developments in veterinary medicine and surgery allowed even more extreme physical characteristics to be bred. Caesarean section has become almost passé and this has allowed breed characteristics to be bred for which make normal birth more difficult. Dystocia probably averages less than 5% in most breeds (Linde-Forsberg, 2001) but brachiocephalic breeds have a lot of problems when whelping. In one study, caesarean section was required in 62% of Boston Terriers, 43% of French Bulldogs and 21% of Boxers (Linde-Forsberg, 2001). The most common cause of dystocia in the bitch was uterine inertia, and some breeds such as the Boxer (28% dystocia) and Smooth-haired Dachshund (10%) seemed particularly prone (Darvelid & Linde-Forsberg, 1994). The skull of the Bulldog should be very large (Anonymous, 1998a) and a large fetal head size may predispose bitches to dystocia. McGreevy and Nicholas (1999) argued that many traits which are really defects have been included in breed standards. These include the fine legs of the Miniature Poodle and Italian Greyhound which are susceptible to fracture if these dogs jump. Defects are not always physical; some lines of full-coloured Cocker Spaniels have had severe aggression problems (Podberscek & Serpell, 1997).

Selection for juvenile physical characteristics, paedomorphosis, appears to reduce the signalling ability of those breeds of dog least similar to the wolf (Goodwin et al., 1997). These signals are used by wolves to reduce the escalation of aggression but as dogs have a higher threshold for aggression than wolves they may not be so important in the dog; however, the significance of losing so many signals is unknown. In addition, many breeds may not be able to use many of their normal signals due to selection for hair cover and type, ear carriage and tail shape. This inability must have some significance in dog-to-human and dog-to-dog communication, but this has not been clarified. It was observed that cutting the hair covering the eyes of the Puli can improve its temperament (Houpt, 1991).

Breed standards have changed over the years and now it is time to change those that predispose animals to physical problems and disease. Although this is not difficult and could be easily carried out, some breeders would complain about changes in standards. Wise and sympathetic breed experts, in conjunction with veterinary advice, should welcome breed standards that improved the health and wellbeing of their beloved breeds.

4.   INHERITED DISEASES

There are more than 370 diseases in dogs that are inherited or have a major hereditary component (Patterson, 2000) and more are identified every year. Many of these diseases show high prevalence in some breeds (Padgett,

1998) but may not be seen in crossbreds because of the masking effect of heterozygosity in modifier genes, or because of co-selection with breed type (Brooks & Sargan, 2001). This myriad of diseases and deformities have increased the veterinary literature and according to Ott (1996) become a cash-cow for the pet-repair type of veterinary practice. The canine genome has accumulated many mutations since domestication. Some have been exploited to develop new types or breeds of dog and may or may not have affected the health of the dog. Others are significant in that they cause disease.

The development of breeds of dogs in the last 150 years required inbreeding and genetic isolation. Individual, valuable champions reduced the gene pool further by being widely used as popular sires. Many breeds had only a few, very important animals which were used widely when the breed was established. More recently few dogs were used as particular traits were sought after (Brooks & Sargan, 2001). In the Netherlands during the last 30 years, only 3–5% of registered dogs were used to produce the purebred dogs present in 1998 (Ubbink et al., 1992). This tendency to use only a small percentage of available stock resulted in an increase of autosomal recessive and other types of inherited diseases in purebred dogs. Selection for particular characteristics may have even increased the prevalence of inherited diseases (Brooks & Sargan, 2001). Inherited diseases were not deliberately added to the dogs' genotype but accompanied the methods of breeding used to develop breeds and winning lines of show dogs (Ott, 1996).

There are four main modes of inheritance of disease (Table 5) and knowing how a particular disease is inherited is fundamental to developing a control programme for that disease in the dog. A disease is suspected of having a genetic background if a higher than expected incidence is found in a particular breed of dog; pedigrees can then be examined to determine whether the pattern of disease fits one of the four modes of inheritance. More than 200 of the inherited diseases of dogs are thought to be simple Mendelian (monogenetic) and 70% of these are inherited from autosomal recessive genes (Patterson, 2000). However, Oberbauer and Sampson (2001) suspected that with some diseases an insufficient number of dogs had been studied to allow an adequate statistical analysis and evaluation of the causal mode.

Table 5. General characteristics of the four modes of inheritance of diseases in dogs.

Autosomal dominant
The mutant gene is generally found as a heterozygous and not a homozygous state
One parent of affected offspring has the disease
Male and female equally affected
If one parent is heterozygous then on average 50% of offspring are affected

Autosomal recessive
Affected dogs are homozygous for the mutant gene
Both parents of affected offspring are heterozygous carriers
The condition may skip generations if a heterozygous is mated to disease-free animals
Male and female equally affected

Sex-linked recessive
Pattern of transmission characteristic (normal females, affected male offspring)
If male and female affected then all offspring affected
Affected males have no affected sons and carrier daughters
Affected males have affected relative on dam's side but not sire's side
On average 50% of male offspring of a heterozygous female will be affected
On average 50% of female offspring of a heterozygous female will be carriers

Polygenic
Condition erratic appearance
Both sexes affected but not necessarily equally
Both sire and dam contribute genes to affected offspring but not necessarily equally
No predictable ratios in pedigrees because number of genes involved not known

---

Adapted from Oberbauer & Sampson (2001)

The incidence of diseases that are transmitted through single autosomal dominant genes is easily reduced as all animals carrying the mutation are affected. Thus, if affected animals are identified and not bred from the incidence will decline. However, there are a few problems with this type of aetiology. The disease may not develop until after sexual maturity or even very late in life and many affected animals may have been bred from and died before the disease was diagnosed. Additionally, some diseases have incomplete penetration and the disease may not be identified. The incidence of these diseases may be reduced by careful examination and selection of breeding stock, and the use of DNA-based tests as they develop (Oberbauer & Sampson, 2001). A reduction in disease caused by polygenic transmission is possible if animals that are severely affected by the disease are excluded from breeding programmes and only those with no disease or minor lesions are used. Hip dysplasia, which is a polygenic condition, is being controlled by restricting breeding to dogs with minor lesions. In Sweden, the prevalence of hip dysplasia decreased during a period when all dogs to be bred had to have a hip score if their offspring were to be registered by the kennel club (Swenson *et al.*, 1997).

Kennel clubs have a major role in encouraging the reduction of inherited disease in pedigree dogs. The Swedish result in reducing hip dysplasia is a good example. By controlling breeding through registration or restricted registration, kennel clubs can force breeders to test their potential breeding stock before they are mated. The types and number of tests will differ between breeds and be influenced by severity of the problem within a breed. Breed inspectors with a good knowledge of the disease, and the physical and behavioural problems of a breed may be appointed to examine stock before breeding and to issue permits to allow breeding, which may reduce disease problems. If a kennel club keeps an open registry with the relevant data relating to the animal's health status available to breeders, then a breeder can know what is available for use as breeding stock. Breeders can, with competent help, reduce the incidence of disease in their animals quite effectively, as shown by the virtual elimination of a storage disease problem in Portuguese Water Dogs (Padgett, 1998).

There are many inherited diseases in dogs that are autosomal recessive in origin. These are more difficult to control, as though animals clinically affected can be excluded from a breeding programme it is difficult to identify heterozygous carriers which remain as a source of the condition. Carrier bitches may be bred only once or twice and not to a carrier dog. Carrier stud dogs are more likely to be identified as they may be used widely and be bred to a carrier bitch. Livestock breeders use test mating to identify carriers but this is generally not acceptable to dog breeders. However, Padgett (1998) recommended test mating of animals that had a high risk of carrying the gene in question if the disease occurred before 2 years of age, on the condition that prospective purchasers were made aware of the risks involved.

Carriers of some diseases have been identified by biochemical tests, but better gene-based tests to identify carriers are required for many of these types of diseases. The development of tests based on DNA allows the identification of pups with particular inherited diseases (van Oost, 1998). DNA tests to identify gene mutations that may cause disease have been developed for more than a dozen diseases (Oberbauer & Sampson, 2001) and linkage-based DNA tests have been developed for some others. The gene mutation responsible for the early onset form of progressive retinal atrophy in Irish Setters has been identified and can be used in diagnosis (Petersen-Jones, 1998). There has been considerable development in this field of characterising inherited diseases at the DNA level, and McGreevy and Nicholas (1999) listed 17 of them. More recently, quantitative genetics has been used to calculate the heritability of ocular diseases in Tibetan

Terriers (Ketteritzsch *et al.*, 2004), and Nicholas and Thomson (2004) were convinced that combining molecular and quantitative genetics would allow breeders to reduce the incidence of multifactorial disorders.

Some breeds such as Greyhounds, selected for specific activities, are relatively free of inherited defects but many other breeds have a large number of them. Within a breed, there may be lines with particular problems and others quite free of them (Ubbink *et al.*, 1998a). Some breeds may have more than 30 inherited diseases but most have perhaps 4–8 that are clinically relevant (Brooks & Sargan, 2001). Moreover, inherited diseases can also be identified in crossbred dogs.

The incidence of disease in some breeds is high. Copper toxicosis was found in 46% and 34% of Bedlington Terriers in the Netherlands and the UK, respectively (Herrtage *et al.*, 1987; Ubbink *et al.*, 2000). Dogs may inherit prenatal and congenital defects, a predisposition to specific neoplasia, or specific diseases (Brooks & Sargan, 2001). For example Kienle *et al.* (1994) found that the Newfoundland, Rottweiler, Boxer and Golden Retriever were at greater risk of having subaortic stenosis, and Padgett *et al.* (1995) found that the hereditability for histiocytosis was 0.298 in Bernese Mountain Dogs.

Many of the rare breeds appear to have fewer diseases than the popular breeds. This may be due to these breeds receiving less veterinary attention than more popular breeds. The incidence of any disease in a local population of dogs is generally not known and even in well-studied diseases such as hip dysplasia, data are only available from those animals subjected to hip scoring. This lack of information makes it difficult for the potential purchaser of an individual pup. Without knowledge of the incidence and its importance in the pedigree of the individual dog, and without the help of someone to interpret those data, all pups are bought to some extent in ignorance.

Padgett (1998) and Brooks and Sargan (2001) compiled comprehensive lists of inherited diseases in dogs which affect the eye; central nervous system; neuromuscular system; cardiovascular defects; haematological defects; renal, hepatic, dermal, enteric, respiratory, reproductive and endocrine disorders; and lysosomal storage diseases. The lists included breeds affected, clinical and pathological signs, and modes of inheritance. Breur *et al.* (2001) listed and discussed inherited orthopaedic problems. Kirk (1986) showed that some breeds including German Shepherd Dog, Cocker Spaniel and English Bulldog had many inherited disorders, but others such as the Greyhound, Pug, and Foxhound had few. Padgett (1998) reported that 40% of Cairn Terriers, 67% of Newfoundlands, 30% of Bichon Frises and 34% of Scottish Terriers had genetic defects and suggested that on average about 40% of all dogs had such defects. Genetic diseases of

Newfoundlands included; hip or elbow dysplasia, undershot or overshot bite, osteochondritis dessicans, patellar luxation, wobbler syndrome, kinked tail; cardiac diseases such as subaortic stenosis, patent ductus arteriosus, and ventricular septal defect; ocular diseases such as entropion, ectropion, everting nictitating membrane, dermoid cyst of the cornea, diamond eye, and medial canthus pocket syndrome; allergic dermatitis, hypothyroidism, umbilical hernia, retained testicle, bloat, allergies, megaoesophagus, trembling, and rage syndrome (Padgett, 1998).

It is obvious that some diseases are more serious than others. Padgett (1998) suggested that breeders should develop a hierarchy of genetic diseases and divide them into serious and mild traits. Serious traits could then be categorised as: (1) painful disorders (e.g. entropion, hip dysplasia), (2) disorders that disfigure or render an animal dysfunctional (e.g. cataracts, progressive retinal atrophy, deafness), (3) lethal disorders (e.g. malignant histiocytosis), (4) disorders requiring treatment for life (e.g. copper toxicosis, epilepsy), (5) disorders requiring surgery (e.g. ventricular septal defects), and (6) disorders that are difficult to control (e.g. subaortic stenosis, osteocondritis dessicans). He categorised mild disorders into: (7) disorders that are easily treated (e.g. hypothyroidism), (8) disorders requiring one-off often cosmetic surgery (e.g. inguinal hernia), and (9) disorders that prevent an animal being used for the purpose for which it was bred (e.g. dentition problems).

Hip dysplasia is one genetic disease that has received a lot of attention in the last three decades. It is a complex condition controlled by the interaction of several genes and environmental factors (Breur *et al.*, 2001). Control programmes for this disease have been established in many countries but reduction in the prevalence of the disease in many populations has generally been disappointing. A reduction has been achieved in controlled populations such as guide dogs by combining individual data with that of related animals (Leighton, 1997), and in Sweden thorough restricted registration (Swenson *et al.*, 1997). However, in the UK (Willis, 1997), USA (Kaneene *et al.*, 1997) and Finland (Leppanen & Saloniemi, 1999) results have been poor. It appears that breeding animals with better-than-average hips will not eliminate the problem and a combination of individual phenotype and that of related animals is needed to identify animals that should be used for breeding (Breur *et al.*, 2001). In the longer term, genetic screening will become available and hopefully assist in reducing the incidence of this disease.

Information about inherited disease in different breeds of dogs is available on several web-sites. The potential purchaser of a pup has the right to know what could go wrong with the breed of dog he/she is interested in. Consumer protection legislation may make the breeder of an animal liable for the veterinary cost and attendant mental anguish caused by the treatment and perhaps euthanasia of a companion animal with a congenital defect which the purchaser should have been warned about. In future, consumer protection legislation may have more effect on the welfare of pedigree dogs than anything else. Breeding programmes are being used to reduce the incidence of some inherited diseases in dogs and more programmes are being developed. Such programmes have to take into account the desire of breeders to maintain the physical characteristics that make the breed unique and their desire to maintain quality in their stock. Accurate pedigrees are a basic requirement of any control programme.

Currently, the control of inherited diseases depends on four major factors: (1) knowledge of how the disease is inherited and whether individual dogs are affected, (2) the motivation of breeders (Padgett, 1998; Swenson, 2001), (3) the actions of organised kennel clubs, and (4) the local legislation. While knowledge of how these diseases are inherited and diagnostic tools continue to improve, the motivation of breeders and their ability to support the identification of affected animals is crucial. Padgett (1998) found that many breeders ceased breeding after a few years due to the problem with pups they sold having inherited diseases, thus veterinary advice through education is important for new breeders.

Genetic counselling to reduce the incidence of hereditary diseases is becoming more important for breeders (Fowler *et al.*, 2000). This type of counselling is based on the breed and individual animal. Where possible, it must allow breeders to utilise good quality stock to maintain the quality of their bloodlines. Oberbauer and Sampson (2001) developed a programme to follow in genetic counselling (Table 6).

National legislation may protect the consumer, that is the purchaser of a pup, from buying defective stock, but animal welfare legislation may also impact on the rights of the individual to breed his/her dog. In Sweden, the law does not allow animals that can pass on inherited diseases to be used for breeding (Swenson, 2001). This law has enhanced the power of kennel clubs to restrict the use of dogs for breeding, and while it may be difficult to enforce it puts pressure on dog breeders to use only animals that are, to the best of their knowledge, not carriers of inherited diseases.

Table 6. Genetic counselling programme for proposed mating of dogs with inheritable diseases.

Step	Strategy
1	Breeder describes goals of breeding programme
2	Identify genetic disease in the breed
3	Outline genetic and clinical tests available for that breed
4	Assemble genotypic and phenotypic data for relatives of dogs
5	Determine if heritability values are available for traits valued by breeder
6	Evaluate valuable and undesirable traits in both animals and their relatives
7	Incorporate all data to calculate outcome of mating
8	If risk of producing pups with genetic disorders outweighs chance of good stock discuss other opportunities
9	Encourage phenotypic and genotypic tests for pups

Adapted from Oberbauer and Sampson (2001)

There are definite breed differences in health (Egenvall *et al.*, 2000ab) and life expectancy (Egenvall *et al.*, 2000c). Some breeds appear to be ill more often than others and to visit the veterinarian more frequently. In Sweden, a number of breeds, including the Boxer, Irish Wolfhound, Great Dane, Doberman Pincher, Giant Schnauzer, Bernese Mountain Dog, Airedale Terrier, Rottweiler, Standard Schnauzer and Flat-coated Retriever, were ranked as having a high morbidity (Egenvall *et al.*, 2000a), while the Siberian Husky, Finnish Spitz, Norwegian Buhund, Schillerstovare, Jamthund, Hamiltonstovare, Grahund, Carelian Beardog, Smalandsstovare, and Norbottenspets had low morbidities. In another Swedish study, life expectancy increased in the following order: Bernese Mountain Dog, Boxer, German Shepherd Dog, Cavalier King Charles Spaniel, Drever, Beagle, mongrel and Poodle (Egenvall *et al.*, 2000c).

5.   BREEDS AND BEHAVIOUR

Only a few breeds of what are now known as lap dogs were originally developed to be companion animals. The majority of breeds were produced and selected to engage in some sort of work or sporting activity. Many are still bred to work and this selection process produces an animal predisposed to engage in particular activities. However, even when not selected for working ability, for several generations many breeds still maintained certain predilections. The Labrador Retriever is predisposed to retrieve and many terriers are predisposed to kill small animals.

A large percentage of owners consider that their dog behaves inappropriately. In one study of 1,422 dog owners, 87% identified one or more behavioural problems with their dogs, and a mean of 4.7 problems per dog (Campbell, 1986b). This may be due to the owners having unrealistic expectations (Overall, 1997) and/or the dogs cannot behave as required. That millions of dogs are killed each year for behavioural reasons is well known. In the selection of working dogs, those that do not perform are killed. Moreover, killing dogs because they do not behave appropriately as companions is probably fundamental to domestication and the development of the dog. The characteristics being selected against remain constant, that is aggression and timidity, but what are new are the characteristics which need to be selected for. In their new form as companions, dogs should be selected for a predisposition to cope with being inactive, isolated, usually living indoors in a city and having no work to do. There are probably few breeders selecting for dogs that are essentially inactive and that require little attention.

In the USA, about 50% of dogs are purebred (Overall, 1997), produced by pedigree breeders. Breeders who breed animals for show characteristics select for a temperament which is meant to represent the breed. Overall (1997) suggested this selection may actually tend to produce greater numbers of aggressive dogs over generations. Moreover, as many of these breeds do not work and are now companions, this direction of behaviour selection is spurious. Selection by breeders should be for dogs that can cope with what companion dogs experience. Dogs can now be divided into different categories based on the reason they were bred and their lifestyles (Table 7). In some breeds (e.g. Labrador Retriever), there may be show lines, working lines and companion lines, but the majority of pedigree pups are produced for show or companion work. There are few breeds registered by a kennel club produced solely for work.

Table 7. Categories of dogs, depending on purpose for breeding and expectations.

Breeding Purpose	Expectation			
	Companion	Showing	Sport	Work
Showing	XXXXX	X	X	
Work	X	X	X	XX
Companion	XX		X	
Haphazard	X		X	

X = relative value and use within breeding group

Currently, the majority of dogs in Europe and North America are owned as companion animals. The continued selection of pets for standard breed behavioural characteristics may have severe welfare implications if these characteristics do not suit the new role. Indeed, some breeds are unsuited to being companion animals and whilst they may be fashionable for a short period of time, their popularity wanes and many are re-homed. The Siberian Husky is an example of a dog unsuited to be a companion dog. This dog is described as independent and having a desire to roam but the understanding owner will find it an enjoyable companion (Anonymous, 1998a). However, the breed society in New Zealand took a firmer stance and suggested that it was not suited to be a companion animal. Backyard or apartment breeders producing pedigree pups for companionship may be producing animals more suited to the market than those breeding for showing. The former may be producing puppies in the environment where the animal will live in the future and may breed their bitch later in life than many show enthusiasts.

Temperament is difficult to define in terms of what to measure in heritability studies but it is generally agreed to be heritable (Willis, 1995).The hereditability of temperament in German Shepherd Dogs, in the USA army was 0.5 (Mackenzie *et al.*, 1985), which suggests that it is easy to select for. Many suggest that owners of companion dogs require a stable, non-aggressive, non-nervous, inactive, easy-going dog (Willis, 1995) that is quiet, obedient and has a low requirement for exercise and activity. These characteristics can be easily selected for and are present in quite a few breeds. However, companion dogs should be able to withstand isolation, not bark much and be tolerant of people. Selecting for these characteristics may not be difficult but it would require a different methodology and focus than those practised by most pedigree breeders. Dogs would not be bred until mature and defined in their behaviour. Temperamental characteristics would take precedence over physical and beauty characteristics.

The existence of breed-specific behaviours was demonstrated by Scott and Fuller (1965) and have been categorised for different breeds by Hart and Hart (1985) and Bradshaw *et al.* (1996) in the USA and UK, respectively. In the UK, breeds were categorised on aggression, reactivity and immaturity. In addition in the American study, breeds were categorised according to trainability but not immaturity. Trainability was not significant in the study from the UK, and may not be relevant to urban living given the limited expectation required of the average pet dog. The exercise requirement of different breeds has not been quantified but breeds with high exercise requirements are obviously not suited to environments where this is unavailable.

In the USA and Japan, five and seven, respectively, of the 10 most popular dog breeds are small dogs. In general their aggression (territorial, dominance, watchdog behaviour, towards other dogs) was average, reactivity (excessive barking, excitability, demand for attention) high, and immaturity (playfulness, destructiveness) low (Table 8). High reactivity might be considered unsuited to urban living but it appears to be acceptable to many owners, at least in a small dog. The size of dog appeared to be important in both countries: 15 and 16 of the top 20 breeds of dog were small in the USA and Japan, respectively. In contrast in the UK, the top breeds were mostly large dogs namely the Labrador Retriever, German Shepherd Dog, Cocker

Spaniel, English Springer Spaniel, Staffordshire Bull Terrier, Golden Retriever, Cavalier King Charles Spaniel, West Highland White Terrier, Boxer, and Border Terrier. High reactivity is not a feature of these breeds except for the Cavalier King Charles Spaniel (Bradshaw *et al.*, 1996). The German Shepherd Dog and Cocker Spaniel, are categorised as highly aggressive and the number of these two breeds is in decline (Table 8).

Table 8. Behavioural characteristics of the 10 most popular breeds of dog in the USA in 2004 and Japan in 2002.

Breed (ranking)	Aggression	Reactivity	Immaturity
Labrador Retriever (USA 1, Japan)	Low	Average	Low
Golden Retriever (USA 2, Japan 9)	Low	Average	Low
German Shepherd Dog (USA 3)	High	Average	Low
Beagle (USA 4)	Average	Average	Average
Dachshund (USA 5, Japan 1)	Average	High	Low
Yorkshire Terrier (USA 6, Japan 6)	Average	High	Low
Boxer (USA 7)	Low	Average	High
Poodle (Miniature) (USA 8, Japan 8)	Average	High	Low
Chihuahua (USA 9, Japan 2)	Average	High	Low
Shih tzu (USA 10, Japan 4)	Average	High	Low
Corgi (Pembroke) (Japan 3)	High	Average	Low
Papillon (Japan 7)	Average	High	Low
Pomeranian (Japan 10)	Average	High	Low

After Bradshaw *et al.*, 1996; Anonymous, 1998a; American and Japanese Kennel Clubs Data bases.

Theoretically, many breeds are unsuited to be companion animals. Bradshaw *et al.* (1996) identified 11 breeds in the UK which were categorised as highly aggressive. Interestingly these breeds have, except for the German Shepherd Dog and Corgis, maintained their popularity (Table 9). Toy breeds with high reactivity (excessive barking, excitability, high demand for affection) but average or low aggression and low immaturity have declined in the UK. Seven of the top 10 breeds in Japan and the USA are in decline in the UK, which suggests that generalisations as to what different societies require in their companion animals cannot easily be made.

Table 9. The number of puppies registered in 1994, 1998 and 2002 in the UK and the classification of breeds according to behavioural characteristics.

Characteristics and breed	No. puppies registered		
	1994	1998	2002
HAg, AR, LI			
Rottweiler	3,070	4,954	5,802
German Shepherd Dog	22,026	20,953	14,177
Doberman	2,183	2,906	2,706
Bull Terrier	1,917	2,523	2,665
HAg, AR, HI			
Parson (Jack) Russell Terrier	345	589	673

Characteristics and breed	No. puppies registered		
	1994	**1998**	**2002**
Corgi (Cardigan)	107	83	56
Corgi (Pembroke)	1,036	627	518
Cocker Spaniel	12,808	14,117	13,417
West Highland White Terrier	14,057	15,131	10,015
Cairn Terrier	2,831	2,297	1,664
Fox Terrier (Smooth-haired)	254	216	167
Fox Terrier (Wire-haired)	737	670	665
Border Collie	2,090	2,245	2,113
AAg, LR, LI			
Bulldog (British)	2,038	2,012	1,936
Chow Chow	801	619	4,07
Great Dane	2,251	2,027	1,736
Airedale Terrier	961	1,018	1,054
AAg, HR, LI			
Toy poodle	2,248	1,729	1,172
Yorkshire Terrier	12,343	8,818	4,222
Chihuahua (Long coat)	1,712	1,214	865
Chihuahua (Smooth-haired)	742	532	373
Miniature Poodle	1,298	1,114	741
Papillon	861	871	642
Miniature Dachshund (all 3)	3,938	3,670	3,164
Pekinese	1,832	1,161	720
Lhasa Apso	3,017	3,358	3,065
Pomeranian	1,242	971	664
Shiz tzu	4,466	4,252	3,113
Standard Dachshund (all 3)	780	889	760
LAg, AR, HI			
English Setter	826	759	568
Irish Setter	1,579	1,443	1,225
English Springer Spaniel	11,904	12,741	12,431
Golden Retriever	14,418	14,803	10,526
Dalmatian	2,794	3,058	2,071
Labrador Retriever	29,118	35,978	35,996
Boxer	8,360	9,612	8,916
LAg, LR, LI			
Basset Hound	1,158	1,161	991
Whippet	1,481	1,619	1,823
English Pointer	868	698	657
LAg, HR, LI			
King Charles Spaniel	280	221	150
Cavalier King Charles Spaniel	13,772	12,702	9,984
Shetland Sheepdog	3,179	2,407	1,500
AAg, AR, AI			
Samoyed	1,270	1,022	523
Standard Poodle	1,391	1,307	976
Rough Collie	3,163	2,339	1,492
Old English Sheepdog	1,505	959	620
Border Terrier	2,766	3,479	5,339

Characteristics and breed	No. puppies registered		
	1994	1998	2002
Beagle	905	939	1,007
Staffordshire Bull Terrier	5,971	9,563	10,711
Scottish Terrier	1,591	1,280	982

Adapted from Bradshaw *et al.*, 1996; UK Kennel Club Data

H = high Ag = aggression (territorial, dominance, watchdog, to other dogs)

A = average R = reactivity (excessive barking, excitability, attention demand)

L = low I = immaturity (playfulness, destructiveness, general activity)

It is important to divide the breeds that work and are also show/companion dogs into strains for both rather than trying to suit every need. Working Parson Jack Russell Terriers are not suited for the average city household any more than working English Pointers are. It is possible that strains of either could be bred to suit city life. It may be best to identify and concentrate on those breeds most suited to urban lifestyle. This may be what is happening as dog owners may be selecting breeds most appropriate to their circumstances. Behaviour suited to the show ring may not be what is required in a companion. Interestingly, some breeds or strains of some breeds of dog have specific behavioural problems. Flank sucking is usually found in Doberman Pinchers while spinning is seen in Bull Terriers (Landsberg *et al.*, 1997).

The breeding of dogs for service roles with the police, military, customs or for guiding the blind has allowed investigation into the heritability of characteristics necessary for success in these fields. Some behaviours like fearfulness are highly heritable (0.4–0.6) and are very easy to breed for. Thus, 'fearful' animals should not be used in breeding programmes (Goddard & Beilharz, 1982). The heritabilities of most behaviours are generally not so high (Schmutz & Schmutz, 1998). There are problems defining specific behavioural and temperamental characteristics and because of this the heritability of 'success' has been quantified. In guide dogs the heritability for success is about 0.44 (Goddard & Beilharz, 1982). The heritability of temperament in military German Shepherd Dogs was 0.50 (Mackenzie *et al.*, 1985). This means that with good selection procedures it is easy to breed dogs with temperaments suited to living as companion animals. This should ensure that fewer dogs would be euthanased or re-homed due to poor behaviour

6.   PREPARATION OF PUPS FOR NEW OWNERS

Breeders have a responsibility to breed healthy and well-behaved dogs and to produce healthy animals fit for whatever purpose they are to be used. As the majority of pups are intentionally bred for companionship, it is up to the breeder to manage these pups in the first 8 weeks of life before sale, in order to maximise their likelihood of success in their new home. This involves three major issues; rearing to maximise confidence and ability, maintaining health, and attempting to match puppies with appropriate owners.

Matching pups to potential owners is difficult and there is little research on the success of matching carried out by many breeders. The behaviour of pups in the first 8 weeks of life does not appear to be related to their future behaviour and the use of attitude tests in pups 7 or 8 weeks of age is probably pointless (Beaudet *et al.*, 1994; Willson & Sundgren, 1998a). It is probably wise to determine whether the living conditions and lifestyle of potential owners suit the type of pup being produced. Different strains of any breed will have particular temperaments and the breeder's knowledge of this is important in selecting homes for their puppies. It appears that size, particularly the size of female pups, affects behaviour. Apparently, large female pups have higher defence and hardness scores as adults (Wilsson & Sundgren, 1998b). The choice of a male or female pup is important as there are significant differences in their behaviour. Females are consistently easier to train (Bradshaw *et al.*, 1996). Potential purchasers of a pup should always see the mother and preferably the father and interact with them to get an idea of what type of dog they are purchasing. This cannot be done in pet shops or shelters and purchasing pups under these conditions is more of a gamble.

Rearing pups to maximise their confidence entails handling them gently on a daily basis. This can start when they are less than 14 days of age and careful handling at this stage will enable them to cope better with stress in the future. Pups 4 weeks of age should be exposed to different elements in the environment. They should meet children and be handled gently by them and meet cats and other animals. They should experience grass, concrete, and linoleum, and be given toys to play with. Gentle restraint at that age allows them to understand a degree of control in gentle circumstances. After 5 weeks of age, they can meet people for longer periods, maybe 30 minutes a day, and be taken outside to toilet after eating. Noises of various types can be introduced. Before pups go to a new home, they should be spending some time alone each day to prepare for the isolation they may meet when they leave the litter.

Most experts recommend that pups be re-homed or sold at about 8 weeks, but if they are not sold, they need more and more environmental experience between 8 and 14 weeks. It is important not to overwhelm young pups with people and other things but it is necessary that they are properly exposed to a wide range of environmental conditions before they leave the litter. Pups reared in large breeding kennels or 'puppy farms' may not get sufficient exposure to the different aspects of the environment to allow them to develop into confident adults. Similarly pups held in pet shops or shelter cages may not be exposed to sufficient stimuli.

Maintaining health by giving anti-parasite tablets every 2 weeks, controlling fleas, and having a suitable vaccination programme are important. Pups should be fed appropriate foods to supplement the bitch's milk and then weaned onto a suitable diet.

The advice that breeders give to the purchasers of their pups is very important and needs to be based on knowledge not tradition. Breeders' advice has to be abreast of advances in knowledge. There are now several quality commercial foodstuffs available for pups of different types and these should be recommended rather than home-made foods or supplements. Supplements to designer diets for fast-growing giant breeds may be damaging.

7.   CONCLUSIONS

The majority of people breeding pedigree dogs do so as a hobby and generally endeavour to produce the best pups possible. It is not a financially lucrative activity. The recognition that some breed conformation standards may lead to physical problems and that some breeds and strains of breeds are not suited to being companion animals is well known. The list of hereditary diseases is lengthening and the methodologies to deal with them are new and developing rapidly. Individual breeders usually limit their production of pups to meet the market. On occasion a fashionable, though unsuitable, breed has been produced to meet a surge in demand. An honest appraisal of the suitability of a fashionable breed as a companion animal is necessary before the problem is exacerbated and dogs are killed or relinquished to shelters or breed-protection groups.

National kennel clubs and breed societies have a responsibility to publicise the physical and behavioural faults of all breeds. However, kennel clubs do not provide quality assurance for purchasers of registered pups. They keep records of pedigrees and run show competitions. Quality assurance is the responsibility of individual breeders. However, it is necessary to change the breed standard of some breeds to reduce the tendency to produce exaggerated characteristics which create health problems. The Swedish Kennel Club has shown how a serious condition such as hip dysplasia can be reduced by limiting registration to pups born to animals that have been assessed. This system changes the attitude of owners and breeders and forces them to assess their breeding stock.

There are several important questions which relate to the welfare of dogs that need to be addressed by all who breed dogs. The number of dog breeders and those who show dogs is declining as problems surrounding the production of pedigree dogs are increasing. It is important to control hereditary health and behavioural problems. It is possible to maintain healthy populations of different breeds of dogs, even those breeds with very small populations, provided the breeding programme is managed correctly. It is important that dogs are selected with characteristics suited to urban lifestyles and for companionship. This means producing dogs with low aggression, and average reactivity and immaturity.

The market may influence welfare issues involved with the production of dogs for the show ring by reducing the numbers produced in several ways. Purchasers of dogs may avoid those breeds with physical or behavioural problems or the likelihood of suffering from one or many hereditary diseases. Breeders may stop producing for the show ring and breed for the companion animal market and thus avoid the different requirements for showing. Breeders will not breed animals for which there is no market and the consumer will influence what breeds continue to be produced. In addition, consumer protection legislation will discourage production of puppies from breeds with serious hereditary problems and force breeders to change breeds or improve them.

Chapter 4

CANINE NUTRITION AND WELFARE

Abstract: The domestic dog is a facultative carnivore with omnivorous potential if
circumstances demand. The dietary requirements of the dog are met by a
quality commercial dog food but many free-ranging dogs are not fed at all.
Poor quality dog foods may cause dietary deficiencies, some of which may be
fatal. Some commercial dog foods may not meet the behavioural requirements
of a dog. A constant diet of dry food is probably boring and dogs may benefit
from variety in their diet including household scraps, large bones and treats.
Obesity is caused by overfeeding and too little exercise and may be the most
important welfare problem of dogs in the post-industrial developed world. In
contrast undernutrition is a serious welfare problem elsewhere. In the last three
decades the development of diets for dogs of different types and at different
physiological stages, and the use of prescription diets to support medical
therapy have improved the welfare of dogs significantly. The latter have led to
exciting improvements in reversing age-related diseases and behaviours of old
dogs. Reducing food intake increases the longevity and health of dogs and
current recommendations may overestimate the energy requirements of adult
dogs. More research needs to be undertaken on the behavioural needs of dogs
in regard to the physical character of their diet.

1.   INTRODUCTION

The Canidae include species that are highly carnivorous, frugivorous or
insectivorous (Ewer, 1973) but most are opportunistic, eating small
mammals, arthropods, fruit and what they are able to find by scavenging
(van Valkenburgh & Koepfli, 1993). Four canids, *viz* African wild dog,
dhole, wolf and the dog, especially the dingo, regularly prey on species
much larger than themselves (Meggitt, 1965; van Valkenburgh & Koepfli,
1993). These species are able to do this because they hunt in pairs or packs

but they also have cranial and dental adaptations suited to killing and consuming large ungulates (van Valkenburgh & Keopfli, 1993).

African wild dogs feed almost exclusively on mammalian prey that they have killed themselves and rarely scavenge, probably to avoid dangerous encounters with lions which predate on them (Creel & Creel, 2002). The dhole is generally considered a pack-hunting carnivore preying on a range of ungulates but also consuming birds, lizards, insects and vegetable matter, but rarely carrion (Sheldon, 1992). Dhole will kill tigers and bears that compete with them for kills. Both African wild dogs and dhole kill by running down and often eviscerating their prey. The wolf is mainly carnivorous, preying on large ungulates (Sheldon, 1992) and smaller prey including rodents, lizards and even fish (Bueler, 1974). It is, however, an opportunist with a penchant for scavenging on rubbish tips, Its diet is quite catholic and includes vegetation, insects, carrion and fresh carcasses. In Finland, wolves kill domestic animals including cats, dogs, horses, cows and sheep (Pulliainen, 1975). Diet is, by necessity, influenced by availability and ability. The availability of food is influenced by location, resident fauna, season and other environmental factors, while ability depends on the size of the hunting group, age, health and experience. Several wolves are required to predate on large ungulates, whilst individuals can catch and kill small herbivores and scavenge effectively.

In general, opportunistic feeding behaviour characterises the canids and this is illustrated by the diet of the dingo. This includes a variety of marsupials such as wallabies and kangaroos, and also sheep, cattle, rabbits, rats, mice, insects, reptiles, carrion and vegetable matter (Meggitt, 1965; Bueler, 1974; Corbett & Newsome, 1975). In Zimbabwe, free-ranging dogs are not effective predators of large ungulates (Butler et al., 2004). In Italy free-ranging and feral dogs did not kill livestock (Boitani et al., 1995; Macdonald & Carr, 1995), although worldwide they frequently kill small livestock, especially sheep. Hunting dogs are often assisted by humans in killing wild pigs or deer, but can kill smaller game on their own. Family dogs often kill sheep, working both as individuals and in pairs or packs. Free-ranging dogs are mainly scavengers, feeding on waste and garbage (Beck, 1975). They do not refrain from killing and eating pups and injured dogs and eating the carcasses of dead dogs. Dogs will dig up a dead dog and consume it. This is a problem when barbiturates are used for euthanasia as consuming the carcass may cause the death of the consumer. In Europe, coffins may have been used initially to prevent dogs from digging up and eating the dead, and in many societies where coffins are not used, human corpses are either cremated or buried deeply and often in a side compartment at the base of the grave which is then filled in with some large stones to stop dogs and other scavengers getting at the corpse.

Welfare problems related to the feeding behaviour and nutritional management of dogs will be discussed in this chapter. These include scavenging *per se*, anorexia, pica, malnutrition, obesity and specific nutrient deficiencies. The widespread availability of quality commercial dog foods, including special feeds for growing puppies and prescription foods for dogs with specific diseases, has led to a great improvement in nutrition in the last three decades. However, there are concerns about the behavioural consequences of these dry and concentrated diets.

2.   FEEDING BEHAVIOUR PROBLEMS

Scavenging is such an important part of dogs' feeding repertoire that, given the opportunity, they will turn over garbage bins and tear into rubbish bags. This is a major social problem of stray and free-ranging dogs. Scavenging is one of the reasons why dogs are not allowed to roam free in most developed societies. When faced with a choice of garbage odours, dogs prefer fresh to aged meat, and meat to vegetables (Beaver *et al.*, 1992). Scavenging of baits containing rabies vaccine may be the most effective way of preventing rabies in free-ranging dogs (Meslin *et al.*, 2000), and chicken heads have been found to be particularly attractive bait for use in such programmes (Kharmachi *et al.*, 1992).

Scavenging can be lethal as the food consumed may be poisonous. Dogs are regularly poisoned accidentally in New Zealand by feeding on possums poisoned by sodium fluoroacetate (Meenken & Booth, 1997), but deliberate poisoning of dogs as part of dog control programmes is performed in many countries. Scavenging is also dangerous because of the possibility of ingesting items such as baited fishhooks or plastic bags which have contained meat. Veterinarians regularly remove objects, including wooden and metal objects, needles, fishhooks and, of course, bones, from the oesophagus of dogs. These items lodge at the thoracic inlet, at the base of the heart or in the caudal oesophagus cranial to the diaphragm. Perforation of the oesophagus may cause localised abscessation, pleuritis, pyothorax or infection at other sites. Foreign bodies are usually removed by forceps via endoscopy. It is important not to damage the oesophagus and sometimes, the object may have to be pushed into the stomach and removed from there. When surgery has to be performed to repair a perforated oesophagus the prognosis is poor.

Dogs may eat materials that their owners consider unacceptable. These include faeces (coprophagy) from dogs, cats and other species, especially herbivores. The latter is so common in dogs that it may be considered normal behaviour (Houpt, 1991; Voith, 1994), although others suggest a medical component to the behaviour (Landsberg *et al.*, 1997). Coprophagy occurs in well-fed dogs with no evidence of gastrointestinal disease. The faeces of different species may be nutritionally valuable, but coprophagy in dogs fed a balanced diet may reflect boredom with their regular diet and a desire for a varied gastronomic experience. Dogs clean up human faeces in many developing countries (van Heerden, 1989) and are seen eating out of discarded nappies. However, soiled nappies were not selected by dogs given alternatives (Beaver *et al.*, 1992). Farm dogs eating the faeces of calves with diarrhoea may develop infectious gastroenteritis, and eating excessive horse manure may cause gastroenteritis. Coprophagy may be one way in which pups ingest appropriate gastrointestinal flora. Bitches stimulate defaecation and eat their pups' faeces during the first few weeks of their lives. The eating of dog faeces by other dogs is probably abnormal, but is unlikely to be significant as the eggs of the endoparasites of dogs generally need some time to develop into the infectious state (Soulsby, 1971).

Pica, an abnormal craving for and ingestion of non-food substances, may have a physiological or psychological basis. Eating grass is a common behaviour of dogs but the reason for it remains unknown. Some have suggested that it may act as an emetic (Voith, 1994) and others that it may have some nutritional value. Some dogs become obsessive chewers and/or swallowers of stones, a compulsive behavioural disorder. The former results in damaged teeth and the latter has to be treated surgically.

Dogs may become anorexic for medical or behavioural reasons. Feeding for the dog may be a social event and the subordinate dog may not feel free to eat in the company of its superiors for fear of causing offence and being attacked. These subordinate animals need to be fed alone is a quiet place. Some dogs feed only in the company of their owners because the dogs need the security provided by their owner's presence. Individual guard dogs may not eat while on duty and need to be fed after work (Neville, 1997). Dogs may also become fastidious feeders by learning that if they do not eat their normal food they will be fed more palatable food by their owners. However, some dogs are poor feeders for unknown reasons (Landsberg *et al.*, 1997) and have to be encouraged to eat by using highly palatable foods (Houpt, 1991).

3.   DOG FOOD

For the greater part of the human-dog relationship, the dog has fed on food scraps and excrement. The nutritional value of household scraps and waste material is often marginal, as was found when domestic dingoes were compared with wild ones (Meggitt, 1965). Many dogs are encouraged to stray to supplement their home diet. In the past only the dogs belonging to wealthy people were likely to have been fed well. Urbanisation of the dog population in industrialised nations in the 19th and 20th Centuries, coupled with increased control of stray dogs, prevented responsible dog owners from allowing their dogs to roam to supplement their home-based diet. This increased pressure on owners to provide a complete diet for their dogs led to the birth of the dog food industry. Initially, the dog food business was a side product of the local butchery, but in the 20th Century dog food production became industrialised and several multinational companies have emerged that control a large percentage of the world market.

There have been major improvements in the nutrition of dogs, and therefore their welfare, as a result of this development. Large companies can afford to undertake research and development of products. They provide diets especially designed for dogs of particular breeds, ages and in particular physiological states (growing rapidly, pregnant, ageing) (Burger & Thompson, 1994). One of the major advances in canine nutrition has been the development of special diets to optimise feeding at different stages of life, and to support the prevention and treatment of a range of diseases. Dog food is available for growing pups, pups of large and giant breeds, adults of giant breeds, adults, active dogs, and old dogs. In addition, diets are available for dogs with a sensitive skin or stomach, obese dogs and for dental care. Diets are available also for older dogs with cognitive dysfunction syndrome, chronic kidney and liver disease, heart problems, urolithiasis, cancer, allergy and gastrointestinal problems. This range of diets allows owners to choose an appropriate and adequate diet for their dogs regardless of age and medical condition. In contrast, poor quality pet foods produced by less knowledgeable companies, or home-made foods may result in deficiencies in specific minerals or vitamins, and sometimes poisoning (Worth et al., 1997). The nutritional requirements of dogs have been defined accurately in the last few decades and are provided for in commercial diets. These diets also appeal to dog owners as they are highly digestible, and faecal volume is small, solid and not too smelly.

Commercial diets may not give a dog an adequate feeding experience, as they are either dry or if wet, have a very soft texture. There is a wide choice of formats in dry kibble, biscuit, canned and rolled foodstuffs, and a great variety of flavours. The food companies have identified what is most palatable to dogs and have designed foods to meet those requirements. Dogs prefer beef and pork to leaner meats, and chicken and lamb are preferred to horsemeat (Houpt, 1996).

The consumption of commercially prepared dog foods has increased dramatically in the last four decades. In the UK, in 1960 49% of households fed commercial dog foods. This had increased to 69% of households in 1994

(Watson, 1996). Apparently 90% of the calories fed to pets in North America, Japan, Australia, New Zealand and Northern Europe, and 30–50% of the calories fed to pets in Latin America, the Pacific rim and the remainder of Europe is derived from commercial pet food (Crane *et al.*, 2000). The quality of commercial dog food varies but quality is guaranteed if the food meets the standards of the Association of American Feed Control Officials (AAFCO). The increase in the feeding of commercial dog food has improved the welfare of dogs (Watson, 1996). However, some suggest the opposite and Lonsdale (2001) considered that commercial dog food was a major factor in the development of oral disease in dogs, and encouraged the feeding of natural foods. The latter include carcasses, either part or whole, of poultry, farm animals and rodents, offal and bones. He suggested that such a diet resulted in healthy gums and healthier dogs.

Some argue against the use of commercial dog foods because many are cereal-based, heat-treated, and contains additives. Others suggest that dogs are carnivores and should be fed a diet suited to this type of animal, namely, dead rabbits or chickens, chicken wings and bones. To them the feeding of grain based diets to dogs is abnormal. Lonsdale (1991) argues that feeding commercial dog foods causes gingivitis and that this condition is the basis of much of the ill-health seen in dogs today. However, meat alone is not a balanced diet for dogs. Some aspects of these contentions are appealing, but it is generally not feasible to feed dead animals to dogs living in an apartment block in New York City or Tokyo. The case about gingivitis is unproven, and while there is a widely held belief that manufactured dog foods have deleterious effects on the health of gums and the condition of teeth, hounds routinely fed raw carcasses had signs of periodontal disease and a high percentage of broken teeth (Robinson & Gorrell, 1997). Dogs are not exclusively carnivorous. They will live off a wide range of foodstuffs. Dogs fed a dental prescription diet had significantly lower scores for accumulation of plaque and gingival inflammation than did dogs fed a regular dry food (Logan *et al.*, 2000). However, dry food may have no advantage over wet food (Harvey *et al.*, 1996) at maintaining periodontal health.

Although, raw-food diets are often recommended for dogs, but there have been few attempts to evaluate these diets (Freeman & Michel, 2001) and none to compare them with commercial dog foods in long-term feeding trials. In one trial, three raw-food diets and two commercial foods were compared. All five diets failed to meet the AAFCO's standards (Freeman & Michel, 2001), having deficiencies or excesses of nutrients that could cause health problems if used long term. There are no long-term feeding trials to validate the suggestion that raw-food diets may cause health problems after prolonged use. Discussion about the effect of raw and commercial diets on canine health will continue until long term feeding trials provide the answers. Raw diets often contain bones and these carry risks of intestinal obstruction, gastrointestinal perforation and broken teeth, although the incidence is unknown. Many people feed commercial diets plus bones. Supporters for raw diets claim that dogs are healthier if fed these diets, but again this is not supported by data from clinical trials. Many people feed dietary supplements to their dogs (Pascoe, 2002) as they feel that commercial diets are inadequate. Again, there is little evidence to support the need for supplements.

A case against the exclusive feeding of dry commercial dog foods is that they are boring. A dog eating dry dog food all its life has a poor repertoire of gastronomic experiences which is probably contraindicated in an animal that is generally omnivorous. Many dogs may engage in coprophagia and pica because of their boring diet. Most veterinarians and dog food companies recognise this deficiency and recommend that between 10 and 25% of the diet can be made up of household scraps and other treats. In Europe, many dog owners feed home-made food to their dogs, but in the UK pets receive a large percentage of their calories as commercial pet food. Remillard et al. (2000) found that in Italy, Germany, France and the UK, 90%, 70%, 60% and 50%, respectively, of calories were provided to dogs from home-made foods.

Of possible welfare significance is the introduction of highly palatable quality diets which combined with lack of activity has led to the levels of obesity seen in dogs today (Sloth, 1992).

4.   MALNUTRITION

Malnutrition is defined as any disorder of nutrition resulting from an inadequate or unbalanced diet. It includes not only acute starvation and chronic undernutrition but also obesity. In dogs, the developmental orthopaedic disease seen in giant and large-breed pups due to excess calcium and energy intake is also malnutrition, as is thiamine deficiency due to intake of poor quality commercial diets. The correct feeding of dogs requires diets to suit the animal's stage in life. Breed also influences dietary requirements, which differ from giant to small breeds at differing stages in life. Dogs engaged in active work or sports require diets which are different from those suited to inactive dogs (Table 1). When the diet is not suited to the individual animal's stage of life and activity level it can then cause malnutrition.

Members of the holistic movement often recommend the feeding of vegetarian, vegan or special diets to their dogs. These diets are usually made at home. Vegetarian diets can be balanced using egg and milk products, but vegan diets may lack arginine, lysine, methionine, taurine, tryptophan, iron, calcium, zinc, vitamin A and some B vitamins (Remillard *et al.*, 2000). Moreover, many veterinary nutritionists warn that homemade food, table food, vegetarian and vegan, and single food diets are more likely to result in subclinical nutritional deficiencies than are commercial diets produced for the stage of life of the dog (Remillard *et al.*, 2000).

Table 1. Feeding requirements for canine athletes.

Athlete	Feeding requirement
Sprint	High quality, digestible, high carbohydrate, low fat, medium protein. Carbohydrate snack 30 minutes after racing to enhance glycogen repletion
Intermediate: working dog	Highly digestible, moderate carbohydrate, moderate to high fat, moderate protein. Energy density must be high enough to allow intake of daily energy requirements. Feed after exercise or 4 hours before work. Snacks can be given during work
Intermediate: in training	As for working dogs, but allow 6 weeks to change over to a working diet if the dog works seasonally
Intermediate: idle	Feed as for normal adult dog
Endurance	Highly digestible, high fat, low carbohydrate, moderate protein, in an energy-dense format. Feed after, or more than 4 hours before, racing and training

Adapted from Toll & Reynolds (2000)

A physical examination will help define a dog's nutritional status, but knowledge of the dog's history is important when recommending adjustments to its diet. Most dogs in average condition should have a waist behind the thorax when viewed from above, and a slight abdominal tuck when viewed from the side. Of importance is bodyweight as a percentage of expected weight for a dog of that breed or type and size. The average bodyweight and height at withers for adult males and females of most of the different breeds (Table 2) are available (Debraekeleer, 2000). Average bodyweights are not available for crossbreeds. A dog's body condition score (BCS) will give an idea of its fat stores and, perhaps, muscle mass (Figure 1). This is achieved by assessing subcutaneous fat, especially over the ribcage, down the topline, at the tail base and ventrally along the abdomen. Body condition scoring was developed in the cattle and sheep industries to assess the body condition status of stock at important times for production, such as before mating, and there is a close correlation between BCS and body composition in cattle. The relationship between body composition and BCS has not been well demonstrated in dogs, but it is used widely to complement bodyweight in assessing nutritional status.

There are illustrated body condition scoring sheets for dogs, to facilitate accurate and repeatable scoring. Scoring systems have either 5 or 9 categories. A five-point BCS (Thatcher *et al.*, 2000) is described in Figure 1. In a clinical or shelter context, BCS can be used with relative bodyweight to estimate an animal's nutritional status and monitor changes if a weight increase or reduction programme is undertaken. The relative bodyweight is the dog's actual weight divided by its optimum weight. A dog of normal bodyweight should have a BCS of 3 in a 1–5 scoring range. The ribs should be easily palpated under a light covering of fat. Pelvic bone prominences should also be easily palpated.

Table 2. Canine bodyweight and height at withers, for various breeds of dog.

	Bodyweight (kg)		Height at withers (cm)	
Breed	**Female**	**Male**	**Female**	**Male**
American cocker spaniel	11	12.5	34-36	36-39
Beagle	12	14	32.5	38
Belgian shepherd (all)	28	28	56-60	60-65
Border terrier	5-6.4	6-7	25	25
Boxer	24	32	53-59	56-63
Bulldog	18-23	23-25	Na	Na
Bull terrier	23.5	28	52.5	55
Cav. King Charles spaniel	5	8	30	33
Chihuahua	≤2.7	≤2.7	16	20
Dachshund (miniature, UK)	4.5	4.5	Na	Na
Dachshund (miniature, USA)	≤5	≤5	Na	Na
Dachshund (standard, UK)	9	12	Na	Na
Dachshund (standard, USA)	7.3	14.5	Na	Na
Dalmatian	22.7	27	47.4	57.5
Doberman pinscher	29	40	60-65	60-70
Fox terrier (smooth and wire)	6.8-7.7	7-8.2	≤39	≤39
German shepherd dog	32	43	55-60	60-65
German short-haired pointer	20.5-27	25-32	52.5-57.5	57.5-62.5
Golden retriever	25-29.5	29.5-34	50-56	57.5-60
Irish setter	27.2	31.7	57.5-62.5	67.5
Labrador retriever	25-32	29.5-36	54-59	56-61
Lowchen	2	4	20	35
Maltese	1.8	2.7	25	25
Pekingese	3-5	3.6-6.5	Na	Na
Pointer	20-29.5	25-34	57.5-65	62.5-70
Poodle (miniature)	5	5	>25	37.5
Pug	6.5	8	25	30
Rhodesian ridgeback	32	38.5	60-65	62.5-67.5
Rottweiler	40	50	55-62.5	60-67.5
Samoyed	17-25	20-30	47.5-53	53-59
Schnauzer (standard)	15	18	45-46	46-50
Shar-pei	18	25	45	50
Weimaraner	32	38	57.5-62.5	62.5-67.5
Welsh corgi (Pembroke)	10-12.7	10-13.6	25	30
West Highland terrier	7	10	25	27.5
Yorkshire terrier	≤3.5	≤3.5	22.5	22.5

Na = not available. Adapted from Debraekeleeer (2000)

Figure 1. Body condition scoring sheet for dogs (adapted from Thatcher *et al.*, (2000)).

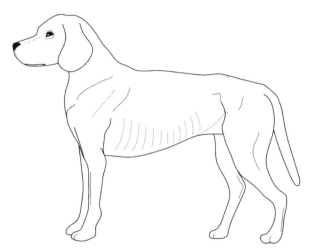

Condition Score 1: Ribs are easily seen and the tail base is obvious. All boney structures have no covering of fat and are easily palpated. Dogs over 6 months of age have a severe abdominal tuck and an hourglass shape when viewed from above. All boney protuberances are easily felt, with no overlying fat.

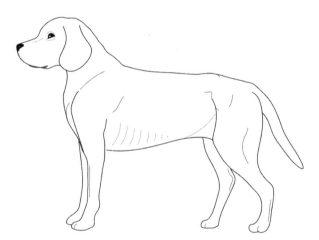

Condition Score 2: Ribs are easily palpable, with slight fat cover. The tail base is a raised structure, with little tissue between skin and bone. The boney protuberances are easily felt, with minimal fat between skin and bone. Dogs over 6 months of age have an abdominal tuck and a marked hourglass shape if viewed from above.

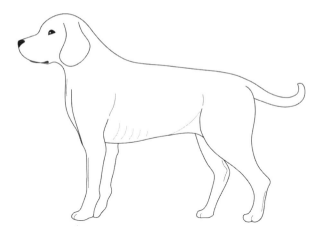

Condition Score 3: Ribs are palpable, with slight fat cover. The tail base has a smooth contour and is thickened. The boney structures are palpable under a thin covering of fat, and boney protuberances are palpable under a small amount of overlying fat. Dogs over 6 months of age have a small abdominal tuck when viewed from the side and a defined waist when viewed from above.

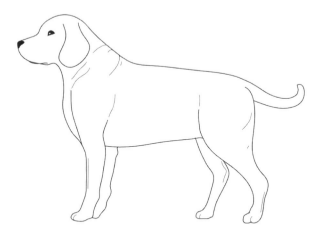

Condition Score 4: The ribs are difficult to feel, with moderate fat cover. The tail base has a moderate amount of tissue between skin and bone. The boney structures are palpable and the prominences are covered with a layer of fat. Dogs over 6 months of age have no abdominal tuck. When viewed from above there is no waist and the back is thickened.

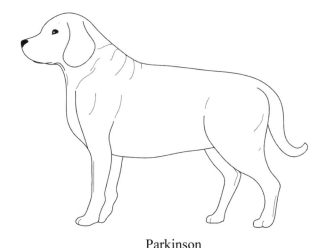

Parkinson

Condition Score 5: The ribs are difficult to feel under a thick layer of fat. The tail base is thickened and difficult to feel under a layer of fat. The boney protuberances are covered with a thick layer of fat. Dogs over 6 months of age have no abdominal tuck and a pendulous ventral bulge. When viewed from above there is no waist and a markedly broadened back.

Dogs that are grossly overweight or underweight are obvious, but the more subtle problems with malnutrition may not be so easily identified unless clinical signs of disease become apparent and are investigated. These include developmental orthopaedic disease, eclampsia, weight loss (especially during lactation), and poor performance or diarrhoea during work or sport. Dogs that are working hard require food with sufficient energy density to allow for their high daily energy requirement. If they are underfed then performance will suffer and they will lose weight rapidly (Toll, 2000abc).

4.1   Starvation

Hungry, emaciated dogs are common in many countries. These animals may be owned or not, free-ranging or feral. Many stray animals are allowed to wander in order to scavenge and feed themselves, but are fed a proportion of their diet by their owners. In one study in Kenya, 95% of dogs were fed household scraps, although only 19% were house-bound (Kitala *et al.*, 1993), while in Zimbabwe few dogs were fed by their owners every day

(Brooks, 1990). In Harare, about 10% of dog owners bought dog food and dogs were usually fed maize meal and scraps (Hill, 1985). House-bound dogs fed household scraps may be underfed and suffer malnutrition and might benefit from being allowed to scavenge. Feral and free-ranging dogs often survive by scavenging from garbage cans and on dumps (Boitani *et al.*, 1995; Macdonald & Carr, 1995). Starvation may be acute, with virtually no food intake over a period of days due to illness, poverty or neglect, or chronic, with inadequate food intake over a period of time. Starvation in an otherwise healthy dog will result in an initial loss of body fat followed by loss of muscle. Starvation leads to immunosuppression and leaves the animal susceptible to infections; it eventually leads to death. The point at which a dog is undernourished is difficult to define, but lies somewhere between a BCS of 1 and 2, although it cannot be defined clinically. This may lead to difficulties when trying to establish whether or not a dog is being underfed.

Many athletic dogs are fed well but maintained with a very low cover of body fat. Greyhounds have more muscle (58% of body mass), less fat and the same amount of bone as other breeds (Gunn, 1978), while endurance racing dogs such as huskies have probably more fat and less muscle. The feeding of hard-working dogs is particularly important. Sheep and cattle dogs and hunting dogs may be very active and need a high-density calorific intake. If fed insufficiently these dogs may lose weight and not be able to work effectively. Maintaining the balance between adequate feed intake and optimal weight is difficult. Hard-working dogs do not need to carry excess fat and are hindered by it.

Simple starvation occurs when dogs are deprived of sufficient food and do not suffer concurrent disease. It usually results from neglect and is not an uncommon reason for dogs being seized from their owners by animal welfare personnel. Prosecutions occur generally when the dog is in an extreme state, as this can be defined by bodyweight and BCS. In the early stages of starvation, animals change from using a mixture of energy sources to using primarily fatty acids. Carbohydrate metabolism is changed profoundly during the first week of acute starvation. Dogs maintain glucose levels during the first two days of starvation, through glycogenolysis and gluconeogenesis. By Day 3, there is a reduction in metabolic rate which continues for weeks to slow fat and muscle catabolism in an effort to survive long-term starvation. The liver releases ketone bodies from fatty acids within the first few days as an energy source for non-glucose-dependent tissues. Fat becomes an important fuel source after 3 to 5 days and protein catabolism also becomes important (Remillard *et al.*, 2000).

During the refeeding of starved dogs, the diet should initially be predominantly fat and protein and should change from simple to complex ingredients (Donoghue & Kronfeld, 1994). Feeding large amounts of carbohydrates must be avoided during initial refeeding as an excess may cause metabolic problems. Initially, meals should be small and offered several times each day.

4.2   Obesity

Obesity is defined as the excess accumulation of fat, and while this is associated with being overweight, dogs that are overweight may be suffering from ascites. Weight can be considered normal if it is 1–9% above optimum weight, overweight if 10–19% above optimum weight, and obese if more than 21% above optimum weight (Burkholder & Toll, 2000). The optimum weights of crossbreds are not available and this scale may be difficult to use clinically (Markwell *et al.*, 1994). Fat mass can also be used to define obesity. Dogs in optimum condition have 15–20% of their total bodyweight as body fat, and when it becomes >20–30% the dog can be considered obese (Burkholder & Toll, 2000). Determining fat mass may be carried out using morphometric measurements. The pelvic circumference is proportional to the amount of fat in dogs and can be used with other measurements, such as hock-to-stifle length in a formula to determine body fat (Burkholder & Toll, 2000). There are several laboratory-based methods of measuring body composition in dogs, including densitometry, total body potassium, X-ray absorptiometry, computed tomography, and magnetic resonance imaging, but these are not useful for clinical examination of obesity in dogs (Markwell & Butterwick, 1994). However, ultrasound may be used to measure subcutaneous fat and is a method by which obesity may be quantified reliably (Wilkinson & McEwan, 1991).

Obesity is the most important form of malnutrition in dogs in the developed world, and studies in the UK, USA and Europe have shown that more than 25% of dogs are overweight or obese (Table 3). As such, obesity is one of the major welfare issues for dogs in these countries as it has a detrimental effect on the longevity, quality of life and health of the affected animals. In one of the largest studies of obesity in dogs, 74% of 8,268 dogs had an acceptable bodyweight (Edney & Smith, 1986).

A large number of diseases are associated with obesity. These include cardiovascular, pulmonary and dermatological disorders, diabetes mellitus, lower resistance to infectious diseases, and articulo-locomotor problems (Sloth, 1992). Dogs found by Edney and Smith (1986) to be grossly obese were more likely to have circulatory, articular and locomotor problems, but there was no relationship between obesity and cutaneous, reproductive or neoplastic conditions. Obese dogs may have reduced exercise and heat tolerance (Sloth, 1992) which may affect the quality of life of dogs thought to have high exercise requirements and/or living in warmer climates.

Table 3. Incidence of overweight, obese and grossly obese dogs in different countries.

Country	Number of dogs	% Overweight	% Obese	% Grossly obese	Reference
Australia	657	25			Robertson, 2003
Austria			44		Steininger, 1981
UK			33		Anderson, 1973
UK			28		Mason, 1970
UK			30		Sibley, 1984
UK	8,268		21.4	2.9	Edney & Smith, 1986

Obesity develops when dogs are in a positive energy balance for a long period of time. Many animals can maintain a balance between intake and activity, and remain at, or near, their optimum weight. Some animals do not control weight gain and this may be due to a number of factors including genotype, de-sexing, age, activity and food type and human behaviour. De-sexed females are twice as likely to be obese as entire females, and de-sexed males also tend to become obese (Edney & Smith, 1986). Owners may not realise that the manufacturer's recommendations are guidelines to be adjusted for each individual dog (Sloth, 1992). In an Australian study, overweight dogs were likely to be neutered, fed snacks, fed once daily, and living in a single-dog household. The odds of obesity increased for each year of life and decreased for every hour of exercise weekly (Robertson, 2003). Owners of obese dogs were more likely to have the dogs in bed with them, and talk to them, and were more likely to spend time watching their dogs eat than were the owners of non-obese dogs. Edney and Smith (1986) found that certain breeds (Long-haired Dachshund, Labrador Retriever, Cairn Terrier, Cocker Spaniel, Shetland Sheepdog, Basset Hound, Cavalier King Charles Spaniel, and Beagle) were prone to obesity, whereas other breeds (German Shepherd Dog, Greyhound, Yorkshire Terrier, Doberman Pincher, Staffordshire Terrier, Whippet, Lurcher) were not. The type of diet did not influence the incidence of obesity (Edney & Smith, 1986) but Sloth (1992) suggested that dogs fed a home-prepared rather than a commercial diet were more prone to obesity. Owners who were obese (Mason, 1970) and those in older age groups were more likely to own an obese animal. Current recommendations may overestimate the energy requirements of adult dogs (Butterwick & Hawthorne, 1998).

Management of obesity in the dog involves modifying the behaviour of the owners such that they manage the feeding and exercising of the dog more effectively. Good owner compliance is essential and family agreement is necessary if any weight-reducing programme is to work. Norris and Beaver (1993) recommended that clients should keep a diary of the feeding behaviour and activity level of the dog and then treat the obesity by feeding scheduled meals in a fixed location and constantly monitor the food given and activity. When pen-housed dogs were placed on calorie-restricted diets, they were initially more active around feeding time, but this was followed by a decrease in activity. Severe calorie restriction may induce a decrease in activity in penned dogs and thus be less productive if weight loss is required (Crowell-Davis *et al.*, 1995).

If a dog is 15% overweight then an initial target weight loss of 15% should be set. This can be achieved within 12 weeks (Markwell *et al.*, 1990). The diet offered should be 40% of its calculated daily maintenance requirement for its target weight (Sloth, 1992) (Table 4). The food should be fed in several small meals if possible and a gradual increase in activity is recommended. The patient should be weighed every 2 weeks. In a study of controlled calorie reduction, dogs were allowed 209 kJ metabolisable energy per kg (target weight) (0.75) per day, with a target weight 15% less than initial weight. The dogs lost about 1% of weight per week over a 12-week period (Markwell *et al.*, 1994). Dogs are used as models of obesity for research into insulin resistance (Villa *et al.*, 1998) and hypertension (Pelat *et al.*, 2002) in humans.

Table 4. Energy requirements for weight reduction in dogs.

Target bodyweight (kg)	Maintenance energy requirement (kcal)	40% of energy (kcal)
5	420	170
10	705	280
15	955	380
20	1,180	470
25	1,400	560
30	1,600	640
40	1,990	795

Adapted from Sloth (1992)

4.3   Specific forms of malnutrition

Developmental orthopaedic disease includes a group of musculoskeletal disorders in growing dogs, especially large and giant breeds. Hip dysplasia and osteochondrosis are the two significant diseases that are influenced by nutrition. Excess intake of calcium and energy, in conjunction with rapid growth, are important factors in the development of these conditions. It is important to limit food intake and growth of the susceptible breeds. Limiting food intake had a beneficial effect on reducing hip dysplasia in Labrador Retrievers (Kealy *et al.*, 1992).

Thiamine deficiency is uncommon, but is one of the few uncomplicated deficiencies seen in dogs. It may be caused by food being cooked at too high a temperature for a prolonged period of time, or by a diet of fish. It causes anorexia and weakness. Poorly produced commercial diets may cause thiamine deficiency.

5.   DIET AND LONGEVITY

A restricted diet is known to increase longevity in rodents. When one group of Labrador Retrievers were fed 25% less than a second group which was fed *ad libidum* from 8 weeks to 3.25 years of age, and then fed 25% less than the diet fed to the second group to prevent obesity, the dogs fed the lesser amount had a median life span (when 50% of dogs died) significantly greater (13 years) than the group fed more (11.2 years) (Kealy *et al.*, 2002). In addition, their maximum life span (when 90% had died) was 14 years compared to 12.9 years for the group fed more. This difference was not significant, but the sample numbers were small. The dogs fed less were 26% lighter and needed treatment for osteoarthritis later (13.3 years) than did those fed more (10.3 years). The BCS (1 emaciated - 9 severely obese) from 6 to 12 years was lower in the dogs fed less (4.6) than in those fed more (6.7), and Kealy *et al.* (2002) recommended that dogs be fed to maintain a BCS of less than 5 on a scale of 1 (emaciated) to 9 (severely obese).

6.   DIET AND BEHAVIOUR

The effect of diet on the behaviour of dogs is a subject about which little is known. Many veterinarians and dog trainers use diet in their treatment of behavioural problems (Mugford, 1987), particularly aggression and compulsive behaviours, but apart from individual clinical cases little has been demonstrated in clinical trials. When dogs with dominance aggression, territorial aggression or hyperactivity were fed a diet low in protein, only those with territorial aggression based on fear showed a reduction in aggression (Dodman *et al.*, 1996). When tryptophan was added to a low protein diet, owner-derived aggression scores in dogs with territorial or dominance aggression were reduced (de Napoli *et al.*, 2000). The restriction of calorific intake decreased activity in penned dogs and may have increased aggressive behaviour (Crowell-Davis *et al.*, 1995). The addition of antioxidants to diets appeared to modify age-related behavioural changes in the dog, with a reduction in age-related house-soiling, and reversed behaviours seen with cognitive dysfunction syndrome. The addition of a premium diet to a programme of human interaction reduced hypothalamic-pituitary-adrenal axis activity in dogs housed in a public animal shelter (Hennessy *et al.*, 2002).

7.   ENRICHMENT METHODS

The addition of specific types of food such as large raw bones, pigs' ears and treats make eating a more interesting experience, and dogs may display food-related aggression around such materials. This suggests that these are highly valued and indicates that perhaps having such items of food is a behavioural need. However, when extras are fed their nutrient value must be removed from the normal diet. Also, bones should not be fed to rapidly-growing puppies of the large and giant breeds as this will increase their calcium intake substantially and may lead to problems.

When an animal lives in an inadequate environment and is showing signs of abnormal behaviour it is generally recommended that the environment be enriched to allow the animal to engage in a wider range of activities. Making food acquisition more difficult is a standard method of environmental enrichment as is varying the type of food offerred. There is a range of feeding devices available which make the dog work in some way for its food. These include pipes or bones into which sausage or canned food is pushed and which the dog empties by licking the food out, and double-skinned dry food distributors that have to be moved around to get them to drop out kibbles of food. Whether these do much for an individual dog is unclear, but owners like to imagine the dog is working for its dinner. In laboratory dog cages, strips of rawhide have been hung up to keep the dogs occupied.

8.   CONCLUSIONS

The nutrition of owned and valued dogs has improved over the last 50 years as enhanced knowledge of the nutrient requirement of dogs has been used in the preparation of commercially available dog foods. The dietary requirements of dogs at different stages of growth and ageing can be met with specially prepared diets. These help control the incidence of disease. Special prescription-type foods are also available for dogs with specific disease problems. The welfare significance of feeding commercial *versus* homemade diets remains unclear. Obesity and starvation are important welfare problem for many dogs.

Chapter 5

HEALTH AND WELFARE

Abstract: The health of dogs in wealthy, developed countries is generally excellent. The number of veterinarians primarily interested in canine medicine and surgery is high and continues to increase. Knowledge of the diseases of dogs and their prevention and treatment has increased greatly in the last 40 years and continues to grow. This advance in knowledge has allowed veterinarians to support the extreme physical characteristics found in some dogs. It has also allowed veterinarians to prolong the lives of dogs affected with previously incurable diseases. Veterinary surgery is also now highly developed and surgery such as artificial hip replacement is becoming common. Organ transplantation will soon be normal practice. Health insurance for dogs has enabled many clients to avail themselves of what previously would have been prohibitively expensive treatment for their dogs. It has facilitated the development of veterinary practice. The health of the majority of dogs living in underdeveloped countries remains poor and infectious disease epidemics and parasites are very common in these populations.

1.   INTRODUCTION

Knowledge of the diagnosis, treatment and prevention of diseases of dogs has improved dramatically in the last four decades. This has occurred due to a change in the human-dog relationship which has seen the dog become an important companion animal. Livestock farmers are willing to finance disease prevention and the treatment of individual animals if it is economically worthwhile. If treatment is too difficult or expensive and the prognosis is guarded then it may be more economical to kill the animal rather than treat it. The health of a companion animal is perceived differently. The animal is of little economic value but its personal value

101

makes it worthy of treatment. The amount of money that an owner will spend on an animal will vary greatly and the major reason for poor health in dogs in wealthy countries is now neglect and ignorance of owners rather than a lack of veterinary knowledge.

The development of pet health insurance in many countries allows owners to afford treatments that would have been prohibitively expensive in the past. Where pet health insurance is common, veterinarians are able to develop skills and obtain equipment to treat conditions which were not treated in the past due to economic, medical or surgical constraints. This ensures that many dogs that would have had to be killed or allowed to suffer can be treated. To a large extent, developments in canine medicine and surgery have followed developments in human medicine and surgery, although dogs are often used as models for human disease and surgical development. However, there have been developments in the area of canine medicine. In this chapter, the positive and negative effects of veterinary medicine and surgery on the welfare of dogs will be discussed.

2.   THE VETERINARY PROFESSION

Veterinary medicine has an ancient tradition based in farriery and herbalism. An Egyptian fresco of around 1000 BC shows a man taking a calf from a cow, perhaps the earliest illustration of dystocia and its treatment in an animal. Veterinary medicine has a long history in the Near and Middle East and China (Dunlop & Williams, 1996), it is likely that Alexander the Great had veterinarians to look after his horses and that Genghis Khan had a cadre of veterinarians to mind the horses on their great march west.

The great cattle plagues of the 18th Century in Europe stimulated establishment of the first veterinary school in France in 1762 (Wilkinson, 1992). In the 18th and 19th Centuries, the profession was occupied with the health of horses and cattle, although it is probable that dogs were also attended to. Blaine published a book on equine and canine diseases in 1800 and in 1817 a book on canine pathology was published (Dunlop & Williams, 1996). Veterinary knowledge of canine diseases continued to increase and the development of general anaesthesia and radiology allowed surgeons like Hobday in England to improve the diagnosis and treatment of illnesses in dogs (Dunlop & Williams, 1996).

The demise of the horse as a military animal and its replacement as a traction animal by steam-, petrol- and diesel-powered engines in the early part of the 20th Century impacted heavily on the veterinary profession, which turned its attention to farm animals to justify its existence. This occurred especially after 1945 when the emphasis on increased food production allowed it to work in conjunction with agricultural science in increasing livestock and poultry production.

In the last 50 years, the emphasis of veterinary science has changed from livestock, horses, and working and sport dogs to a primary interest in companion animals. In the USA prior to 1950, few veterinarians made a living from small animal medicine, but in 1991 more than 50% of the 60,000 veterinarians in North America described themselves as predominantly small animal veterinarians, and nearly 23,000 were exclusively so (MacKay, 1993). Recent figures from the American Veterinary Medical Association illustrate the trend (Table 1). Predicted changes in veterinary employment for the next decade show this trend continuing and the number of veterinarians in the USA in small animal practice is predicted to rise from 44,667 in 2005 to 52,741 in 2015 (Table 2), despite numbers of dogs levelling off during the last decade and the number of households with dogs declining (Brown & Silverman, 1999). Other branches of veterinary science are expected to remain stable during this period (Table 2).

This continuing increase in the number of small animal veterinarians will intensify competition in the profession and may make veterinary services more affordable (Boivin, 1998). This outcome may be necessary, as in the USA most pet owners were willing to pay only a little over $1,000 to keep their favourite pet from dying (Brown & Silverman, 1999).

Table 1. Membership of the American Veterinary Medical Association (AVMA) and the type of practice they engaged in over the last four decades. Personal communication with Allison Shepherd, AVMA.

Practice	1962	1971	1980	1990	2000
Small animal	3,963	7,050	11,676	21,237	30,987
Mixed (mostly small animal)	NA	NA	4,041	5,308	5,969
Total membership	23,623	25,665	25,102	38,458	50,070

NA = not available

Table 2. Predicted changes in the number and employment of veterinarians in the USA until 2015.

Employment	1997	2000	2005	2010	2015
Small animal	39,875	41,416	44,667	48,415	52,741
Large animal	11,728	11,738	11,951	12,049	12,081
Academia	5,784	5,792	5,829	5,865	5,900
Industry	1,962	2,009	2,152	2,337	2,431
Government	3,986	3,989	4,021	4,049	4,064
Total	63,351	64,944	68,620	72,715	77,317

Adapted from Brown & Silverman (1999)

This change in the direction of the veterinary profession is reflected in the curriculum of most veterinary schools in the developed world, where companion animal medicine and surgery dominate years of clinical training and the dog is the most studied species. Student demography has also changed over the last 30 years from predominantly males with a rural background to females from urban areas (Aitken, 1994; Miller, 1998). This probably reflects a change in the veterinary profession from being concerned with animal production to caring for animals. This should improve the welfare of the dog over the next few decades, as female veterinary students rated themselves as having higher levels of emotional empathy with animals than male students (Paul & Podberscek, 2000). Thus they should be more concerned about the welfare of dogs (Herzog *et al.*, 1991). Men may be perceived as more threatening by dogs than women (Hennesy *et al.,* 1997; Wells & Hepper, 1999). There has also been a change in the use of animals during the teaching of veterinary undergraduates in the last decade. Fewer animals, especially dogs, are used in physiological practicals. The use of unowned dogs in surgery training when the animal is allowed to recover from anaesthesia, has declined greatly.

The increase in the number of veterinary practitioners available for dog owners has been accompanied by an increase in the number of veterinary specialties available for referral consultations. In the 1960s and 1970s, veterinarians developed interests in individual species and in particular aspects of veterinary medicine and surgery. In the USA, this led to the development of specialist groups, and individuals became expert in particular areas of medicine and surgery both within and between species. These groups developed training programmes and interested veterinarians could then study and proceed through a series of examinations to become registered specialists. This development spread worldwide and the Royal College of Veterinary Surgeons in the UK and the Australian College of Veterinary Scientists have specialist registration for individuals with particular knowledge in specific areas, and there is also a European list of veterinary specialists. This ongoing development of specialisations has had a major influence on the progress of clinical knowledge and has resulted in practices with multiple specialists providing referral services for general practitioners. Specialists are now registered in a wide range of disciplines (Table 3) and are found in private specialist practices and veterinary training colleges worldwide.

Veterinarians in companion animal practice do not receive government support and have to be financially self-sufficient. This raises issues between time spent with patients and clients, the fees charged and a client's ability or willingness to pay. Veterinary fees for individual clients can be held low only if sufficient clients use the services and each client takes up a limited amount of time. This causes veterinarians to narrow their focus and concentrate on the current or obvious illness, possibly ignoring the overall wellbeing of the animal (Pascoe, 2002). Veterinarians may not spend time investigating the nutrition, home environment and behaviour of the animal as the economics of practice force them into short consultation times and rapid diagnosis (Pascoe, 2002). This niche in veterinary medicine has been filled to some extent by those who practice complementary and alternative medicine. They fill a requirement for a more holistic approach to the health of animals.

Table 3. Some of the specialist areas defined by the Royal College of Veterinary Surgeons (RCVS) and the Australian College of Veterinary Scientists (ACVSc).

RCVS	ACVSc
Anaesthesia	Anaesthesia and critical care
Cardiology	Animal behaviour
Dentistry	Animal welfare
Dermatology	Canine medicine
Diagnostic imaging	Dermatology
Neurology	Emergency and critical care
Nutrition	Nutrition
Oncology	Oncology
Ophthalmology	Radiology
Pathology	Small animal medicine
Reproduction	Small animal surgery
Small animal medicine	
Small animal surgery	
Small animal surgery (Orthopaedics)	
Small animal surgery (Soft tissue)	

The veterinary profession had widened its area of interest in the last few decades. For instance, veterinarians now regularly advise on pet selection, canine nutrition and behaviour, run socialisation and training classes for young dogs, and engage in genetic counselling. Development of formal training courses for veterinary nurses has had a major impact on the breadth of veterinary involvement in the dog's life and canine veterinary practice is becoming more holistic while remaining bound to evidence-based medicine and surgery. Of particular importance is the growing veterinary interest in canine behavioural problems. This has led to an acceptance that anxiety is fundamental to many of these problems and that alleviation of anxiety is essential in their treatment (Overall, 1997). Consequently, there is a deeper interest in how animals experience life, which is fundamental to animal welfare (Duncan et al., 1993). There is an increased interest in the entire life of a dog from puppy socialisation through training, to the environment that an individual dog occupies and its lifestyle. Veterinarians and veterinary nurses have become involved in the treatment of behavioural problems in dogs (Chapter 10) and in the prevention of such problems (Overall, 1997). In addition, veterinarians are the major source of advice for canine nutrition

(Chapter 3). Practising veterinarians and their staff are well suited to educate clients to improve the welfare of their dogs, and appear to be doing so effectively.

Pet insurance has become common in some European countries such as the UK and Sweden. In the latter, 50% of dogs are insured (Bonnett *et al.*, 1997). However, the rate of uptake of pet insurance is very low in the USA, and in one survey 46% of people were not interested in insuring their pets at all, while 23%, 15%, 8% and 5% were willing to pay $5, $10, $15 or $20 or more per month for pet insurance, respectively (Brown & Silverman, 1999). This lack of insurance may limit development in veterinary practice as dog owners may be unable or unwilling to pay for expensive treatments.

3.   VETERINARY SCIENCE

Veterinary science is an applied science concerned with examining animal health problems and solving them. In the last 40 years, it has undergone major development, as seen by the increase in veterinary books and journals being published. First editions of many of the major texts in canine medicine and surgery were published in the 1970s and they have progressed through several editions. *Canine Medicine and Therapeutics* was first published for the British Small Animal Veterinary Association in 1970 and the 4th edition, edited by N. Gorman, was published in 1998. Ettinger's *Textbook of Veterinary Internal Medicine* was first published in 1975 and is now in its 5th edition. Important textbooks on the diagnosis and treatment of canine behavioural problems have emerged more recently (Overall, 1997). Many veterinary journals began in the 1960s and 1970s but others, such as *Veterinary Dermatology* (1990) and *Progress in Veterinary and Comparative Ophthalmology* (1991), are more recent in origin and reflect the growing specialisation within the profession.

Scientific papers published in these journals have become increasingly concerned with dogs rather than other species. In a survey of the annual index of one journal published monthly, the *Journal of the American Veterinary Medical Association*, entries under canine species increased from 53 in 1960 to 92 in 1970, 71 in 1980 and then decreased back to 53 in 2000, presumably as other journals with a canine focus were created. During this period, entries under bovine declined from 55 in 1960 to 15 in 2000. A similar trend in canine-related papers is seen in the index of the *American Journal of Veterinary Research* but in that journal, papers about cattle also increased.

There have been numerous major developments in the last 50 years in veterinary science, relating to the health and welfare of dogs (Table 4).

Vaccines are now available for all the major infectious diseases, or dogs can be treated reasonably effectively. Many of these developments depended on using dogs as research animals. This has been costly with regard to the welfare compromise that occurred during development and testing. Many treatments, for example improvement in treating diabetes mellitus, have followed advancements in human medicine. Technical developments in canine surgery have occurred alongside similar developments in human surgery, facilitating rapid accession of the techniques by veterinarians. Hip replacement and complex cardiac, gastrointestinal and thoracic surgeries are now common. Organ transplant surgery is not common in dogs yet, and there are concerns about the welfare of donor animals. Intensive care of seriously injured or ill animals and care after surgery have developed significantly in the last decade. Palliative care is still in its infancy in veterinary medicine but will develop as increasing numbers of owners request long-term treatment for dogs with cancers and cognitive dysfunction syndrome.

Table 4. Some major veterinary developments which have improved the welfare of dogs.

- Development of vaccines for rabies, canine distemper, infectious canine hepatitis, leptospirosis, parvovirus disease.
- Widespread use of sulphonamides and penicillin
- Treatment for demodectic mange
- Safe anaesthetics; cheaper, safer and non-narcotic analgesics
- Safe ovariohysterectomy
- Anthelmintics for round worms
- Hip replacement surgery

The major infectious diseases of dogs include canine distemper, infectious canine hepatitis, leptospirosis, rabies, kennel cough and parvovirus disease. Vaccines are available for these and in most wealthy countries the majority of dogs are vaccinated against most of these diseases. Surveys are rarely conducted, but in Ontario more than 95% of dogs had been vaccinated and 88% of rural and 82% of urban dogs had visited a veterinarian in the previous year (Leslie *et al.,* 1994). Nowadays, distemper, leptospirosis and canine infectious hepatitis, which were common canine diseases in most developed countries 30 years ago, are rare. There are concerns about the effect of vaccination on the immunocompetence of vaccinated dogs but there is little evidence to support these concerns, which are offset by the significance and often fatal consequences of the diseases against which the vaccines are used.

Advances in veterinary science resulting in overcoming a new problem was illustrated when parvovirus disease emerged in the 1970s as a new major infectious disease of dogs, particularly young dogs. Treatment protocols were rapidly developed and a vaccine was produced within a couple of years. The quality of the vaccine has since improved and now parvovirus occurs only when pups are not vaccinated against it, usually in poorer societies and countries.

The wide range of antibiotics which are available for the treatment of humans has been adapted for the treatment of dogs and allows more specific targeting of different bacteria. Concerns about the veterinary use of antibiotics influencing development of antibiotic-resistant bacteria in human medicine may result in prohibition of some antibiotics in veterinary medicine in the future.

The majority of internal parasites of dogs are now easily controlled with drugs. Heartworm, a parasite of great concern in dogs in many countries can now be controlled by preventative programmes following advances in veterinary research. Severe parasite conditions of the skin, such as demodectic mange, can now be treated successfully. The availability of effective anthelmintics from both veterinarians and supermarkets has resulted in many dogs now being free of internal parasites which were once commonplace (Bugg *et al.*, 1999). Likewise, the availability of effective, easy-to-apply topical treatments for fleas has made control of this parasite easy.

Veterinarians began using general anaesthesia for dogs in the 19th Century, and early in the 20th Century local anaesthetics became available. There are always risks with general anaesthesia but today it is almost always safe if administered correctly and monitored adequately. The use of analgesics is increasing rapidly due to increased veterinary concern about pain in animals and the development of cheap, safe and effective analgesics (see Chapter 6).

4.   VETERINARIANS AND THE HUMAN-ANIMAL BOND

The human-animal bond is a phrase commonly used in discussing relations between people and their pets and in some cases the human-dog bond may be as complex as the relationship some people have with other people (Stern, 1996). Veterinarians may underestimate the depth of the relationship (Catanzaro, 1988) and the impact that the loss of a dog has on their clients. However, veterinary staff and students agree that compassionate and caring attitudes were necessary during euthanasia and that clients had the right to be present during that procedure if they want to and should be well informed and prepared (Martin *et al.*, 2004). This suggests that veterinarians are sensitive to the needs of their clients, at least during euthanasia of their pets.

In the past, with poorer knowledge and skills, many dogs that were seriously injured, in pain, or sick with diseases which were difficult to treat were subject to euthanasia to end their suffering. Now, with many advances in surgery and medicine, many dogs are treated that would have been killed previously. One of the major animal welfare issues in veterinary medicine is when to treat a dog, which will suffer for some time during convalescence, and when to humanely end its life. This decision is made by the owner, with advice from the veterinarian. Veterinary advice is also important in preventing dogs being surrendered to animal shelters, possibly because it lowers the owners' expectations of what a dog can provide and increase their knowledge of what is normal dog behaviour. The depth of the human-dog bond may have positive and negative effects on the welfare of the dog. It may encourage the owner to pay for the cost of expensive treatment but may prevent euthanasia of a terminally ill dog with a poor quality of life. The veterinarian may gain more financially from treating an animal than from terminating its life, and current emphasis on the human-dog bond supports treatment as the preferred option. In some societies such as in Japan, euthanasia is not as acceptable as it is in western cultures (Kogure & Yamazaki, 1990). Moreover, in western cultures, urbanisation has removed the population from the reality of livestock production, and the development of the animal rights philosophy, has made euthanasia less acceptable to many people. The increase in information on canine health and disease, especially through the worldwide web, has made owners more knowledgeable and more demanding of sophisticated treatment, as has the rise in pet health insurance. While this may appear good for the individual dog, it might mean a life of pain and misery rather than death. It is possible that veterinary science has exceeded its mandate (Lascelles & Main 2002). The question of whether to treat or kill remains pertinent. As an example should we replace the hips of a dog with hip dysplasia or end its life humanely? If a major intent of those concerned with animal welfare is to minimise pain and suffering then it may be more appropriate to kill the animal rather than treat it and have it suffer even for short periods of time.

In 1998, a survey of 478 cases of euthanasia of dogs in veterinary practice found that the majority were for reasons of senility (60%) or terminal illness (27%), while behavioural problems (6%), trauma (5%) and the euthanasia of healthy animals (2%) made up the rest (Edney, 1998). These results suggest that euthanasia is used appropriately, that is primarily to kill old and diseased animals with poor quality of life rather than healthy animals. Assessment of the quality of life of a dog is a subjective process and is difficult to quantify (McMillan, 2000), but specific factors such as feeding behaviour, response to the owner and general demeanour can be useful indicators (Stewart, 1999).

The pain caused by surgical intervention for ongoing disease or chronic conditions such as some cancers, and severe trauma is significant. The surgery may be radical and give rise to severe pain. The affected animal will probably be hyperalgesic and resistant to analgesic therapy (Lascelles & Main, 2002). Unless this is recognised and effectively dealt with using effective analgesic protocols, the severe pain experienced by the dog, may lead an owner to elect euthanasia as a preferable option.

5.   COMPLEMENTARY AND ALTERNATIVE THERAPIES

There is a large and developing interest in complementary and alternative therapies for human and veterinary patients. Some veterinarians have developed interests in homeopathy and acupuncture (Schoen, 2001) and some colleges have specialist chapters in this broad area. However, the majority of therapists are not trained veterinarians.

The effectiveness of alternative therapies as a means of inducing analgesia was reviewed by Pascoe (2002). He found evidence that although acupuncture did cause an increase in nociceptor thresholds in test animals, the clinical evidence that it acted as an analgesic was poor. Veterinarians who recommend nutraceuticals believe that a number are analgesic and there is some evidence of this effect by certain dietary supplements in horses and people. A lot more research needs to be undertaken to demonstrate the analgesic effects of alternative and complementary treatments before they can be recommended to relieve pain in dogs.

6.   CONCLUSIONS

Practicing veterinarians in the developed world now concentrate on the dog as the main species of interest. There have been many significant developments in the ability to prevent and treat canine disease and in surgery. Many of these these developments have followed improvements in human medicine and surgery. Veterinarians may underestimate the human-dog bond but they use euthanasia carefully and usually to relieve the suffering of old dogs with poor quality of life. In many poor countries veterinary care of dogs remains inadequate and dogs still suffer from the major infectious diseases and parasites that are rare or well controlled in wealthier countries.

Chapter 6

PAIN IN DOGS: ITS RECOGNITION AND ALLEVIATION

Abstract: Pain is an unpleasant but important experience. It enables an animal to identify
              and respond to dangerous stimuli and to learn to avoid them in the future. Pain
              is inevitable but in the domestic dog it should be prevented and controlled
              whenever possible. Assessment of pain in the dog is a controversial
              undertaking, comprising two major methodologies. In one method, pain is
              assessed by interpreting canine behaviour using a number of scales. This
              method depends on subjective human evaluation of pain and is influenced by
              individual human perceptions of pain and its expression in dogs. In the other
              method, the behaviours of dogs subjected to painful procedures with and
              without analgesia, are monitored rigorously and compared. The behaviours
              which indicate pain are identified and can then be used to assess the efficacy
              of various analgesic protocols. The use of analgesics in dogs by veterinarians
              is still limited but is growing as safer and cheaper analgesics become
              available. Preventing and alleviating pain is an important way to improve the
              welfare of dogs whn injured, after surgery, or suffering from chronic painful
              diseases.

1.   INTRODUCTION

Pain is defined as an unpleasant sensory or emotional experience associated with actual or potential tissue damage (Merskey, 1979). It is a subjective experience and solipsism, which states that the only consciousness an individual knows is his or her own, proposes that the assessment of pain in another human or species is impossible. It is necessary to reject solipsism before a discussion on the measurement of pain in animals can be undertaken. Bateson's (1991) comment "that though

111

solipsism is logically defensible, most people would treat hard-line solipsism as an absurd stance", is easy to accept. However, everyday solipsism is not hard-line, it is a common factor in human interaction and is more obvious when we try to quantify what animals experience. Indeed, quantifying pain in humans is an elusive and complex undertaking (Richards *et al.*, 1982). The perception of pain and its alleviation differ markedly between individuals (Bhargava, 1994) and are influenced by gender and physiological states such as pregnancy and parturition (Cook, 1997; Machado *et al.*, 1997).

Solipsism has been rejected by analogy and this stance is taken by many, including Singer (1998), who stated "my belief that animals can feel pain is similar to the basis of my belief that my daughter can feel pain. Animals in pain behave in much the same manner as humans do". Although this opinion is questionable it is supported by morphological and physiological studies that demonstrate similarities in the physical mechanisms necessary for pain perception between all mammals (Broom, 2000). Thus, most people believe that mammals feel pain but may disagree about how it is experienced. A second act of faith is required to accept that animals experience pain in a fashion similar to how humans experience it. This is more difficult as individual humans have quite different experiences of pain in response to similar stimuli. However, it is not necessary to believe that animals feel pain as humans do, only to accept that pain is an experience that they will avoid and which will dominate their physiology and behaviour in a similar manner to the experience of pain in humans (Flecknell & Molony, 1997). The great pain physiologist, Patrick Wall (1992), argued that the phrase 'pain in animals' is meaningless as humans do not share identical experiences with animals, but that when we injure an animal we should, by whatever means, assist its return to a normal physiological state.

Today, it is widely accepted that dogs feel pain (Paul-Murphy *et al.*, 2004) and that physiological and behavioural responses can be used to assess and compare the severity of the pain experienced by dogs subjected to different physical insults with or without analgesia. This allows a comparison to be made of the relative unpleasantness of different procedures.

Pain is an important aspect of living and its prevention and alleviation an important aspect of animal welfare. It is useful in that it helps an animal to identify a physically damaging situation, to escape from it, to convalesce, and avoid it in future. Pain is an important component of an animal's experience of suffering. Dogs experience pain during and after injury or surgery, and as a result of some disease states such as cancer, joint disease, periodontal disease and various skin conditions (Taylor, 1985). The assessment of pain and its alleviation are important elements in how we manage dogs. The use of analgesics by veterinarians is still quite low but has increased substantially over the last decade as safer drugs have come on the market. This chapter will briefly describe the physiology of pain, its assessment and the use of analgesics in dogs.

2.   THE PHYSIOLOGY OF PAIN

Pain is the perception of an unpleasant sensation caused by damaging stimuli. The nervous pathway of pain is a three-neuron link, starting in the periphery and projecting to the spinal cord (first-order), ascending the spinal cord (second-order), and projecting to the brain (third-order) (Johnston, 1996). When tissues are damaged nociceptors are stimulated. Nociceptors are part of afferent nerve fibres which respond to physical, chemical or thermal insult. They are numerous in the skin, and internal tissues such as periosteum, joint capsules, arterial walls, muscles and tendons. Most pain impulses are carried in myelinated Aδ and C fibres. The former produces sharp prickly pain and the latter slow burning pain. C fibres are associated with visceral pain as only they are found in the viscera. The threshold of a nociceptor must be exceeded before an impulse travels to the central nervous system. It is similar for all animals. Chemicals such as histamine and bradykinin, which are released by damaged cells, stimulate nociceptors, and other chemicals, such as prostaglandins sensitise nociceptors to stimulation (Sackman, 1991). Inflammation associated with damaged tissues stimulates nociceptors, especially C fibres, and causes a dull aching pain.

Nociceptor afferent fibres enter the spinal cord from the dorsal root and terminate on dorsal horn neurons. Three categories of neurons in the dorsal horn play a role in nociception. Projection neurons relay the nociception to the central nervous system, excitatory interneurons relay nociceptive input to other cells, and inhibitory neurons contribute to the control of nociceptive transmission (Sackman, 1991). The ascending neurons form two different pathways; the ventral spinothalamic tract projects to the thalamus and the ventral spinoreticular tract terminates on the reticular formation in the brain stem. Neurons project from the thalamus to the cortex, where characterisation and location of pain occur.

Tissue damage and inflammation cause pain hypersensitivity and there is an increased response to painful stimuli (hyperalgesia). This may be accompanied by allodynia, when pain is caused by stimuli that normally are not painful (Muir & Woolf, 2001). A 'soup' of substances including contents of damaged cells and sensitizing agents from inflammatory cells may activate sensory nerve endings but also lowers the threshold of high-threshold nociceptors. In addition, spinal neurons become hypersensitised (central sensitisation) to further stimuli and there is an increase in the duration and size of the response to further stimuli. The intensity and character of pain can be reduced by endogenous opioids and biogenic amines in the spinal cord.

Pain is categorised in different ways. In one classification, there are three major categories of pain, namely acute, cancer and chronic pain (Johnson, 1991). Acute pain can be classified as somatic or visceral. Somatic pain arises in the superficial structures (skin, subcutaneous tissues, body wall) and can be divided into first and second pain. First pain is the well-localised initial sharp pain associated with tissue damage, while second pain is the dull diffuse pain which is delayed. Visceral pain arises from thoracic and abdominal viscera and is a feature of serosal irritation. Acute pain is generally easy to treat. Cancer pain has a well-defined onset and is acute, recurrent pain. Pain is considered chronic when it lasts for several months. Its onset is poorly defined and it is often difficult to treat. In another classification, pain is categorised as physiological or pathological (clinical) (Taylor, 2003). Physiological pain is essential for survival. It allows detection of potentially harmful stimuli and reflex responses to them. It hurts, but is of short duration and sensation quickly returns to normal. Pathological pain arises when tissue damage occurs and the injured tissues become hypersensitive (Webb, 2003).

3.   ASSESSMENT OF PAIN

Pain is assessed in animals using two methodologies (Hansen, 2003). The first is animal-orientated and attempts to identify aspects of the animal's behaviour, and physiological, immunological or other responses which can be used to identify pain, and which can be measured and used to assess the severity of pain. The second method uses the human interpretation of behaviour to assess the severity of pain. Both methods have strengths and weaknesses. The former is initially difficult and expensive to undertake both financially and ethically but allows the animal to present the relevant responses. The latter is easier to conduct but does not 'ask' the animal what it is experiencing, only what humans think it is experiencing (Holton et al., 2001), and relies on the subjective evaluation of behaviours whose correlation with other behavioural or physiological indicators of pain and distress have not been confirmed (Hansen, 2003). There are additional problems in that there are individual responses to painful stimuli and some breeds of dogs, for example American Pit Bull Terriers, are considerably more stoical than other breeds.

3.1   Animal-based methods

To determine whether a particular behavioural, physiological, immunological or other response is indicative of pain, a protocol in which sufficient control groups and groups given effective local anaesthetic and/or systemic analgesics before and after treatment must be developed (Mellor *et al.*, 2000). Control groups must include animals receiving no analgesic, hence the high ethical cost (Flecknell & Molony, 1997), and local anaesthetic or systemic analgesics are used to determine their effect on the responses being monitored (Flecknell, 2000). Protocols of this complexity using dogs are uncommon but necessary if the behavioural and physiological responses indicative of pain are to be identified (Table 1) (Fox *et al.*, 1994, 2000). It is preferable if these responses are monitored concurrently or in parallel studies, but behaviours must be identified before developing scoring systems for pain assessment in clinical situations, and they can only be developed by protocols such as those of Fox *et al.* (2000). The pain experienced by animals similar in breed, age, gender and previous experience can be compared using such protocols, but it may not be possible to compare different classes of animals (Stafford & Mellor, 1993).

Table 1. Protocol to identify changes in behaviour following ovariohysterectomy plus different combinations of halothane anaesthesia and butorphanol analgesia.

Treatment		
1.	Control	Catheterised but not anaesthetised
2.	Anaesthesia control	Anaesthetised
3.	Analgesia control	Butorphanol given
4.	Analgesia plus anaesthesia	2 + 3
5.	Anaesthesia plus surgery	2 + Ovariohysterectomy
6.	Analgesia, anaesthesia, surgery	3 + 5

Adapted from Fox *et al.* (2000)

Scientific protocols can be used to determine how an animal responds behaviourally to different painful stimuli, identify which behaviours can be used to identify pain, and to evaluate the severity alleviation of pain. These protocols are of three basic types: (1) observational studies, when animals subjected to surgery (Table 1), or injured or diseased animals are compared with control groups and groups receiving effective analgesia; (2) pain threshold trials, which measure the response of treated and control animals to defined painful pressure, or thermal or electrical stimuli; and (3) choice studies, where animals are allowed to avoid or choose painful or non-painful treatments or to self-medicate with an analgesic to alleviate pain.

Four axioms (Lester *et al.*, 1996) have been used to determine what overt behaviours are useful in identifying pain. They may also be used to evaluate the use of other parameters such as the plasma cortisol response, as indices of pain in any species. The axioms are that a behaviour may: (1) identify pain, if it is seen during and after a tissue-damaging injury but not seen in non-damaged animals; (2) identify nociception and, by inference, pain, if it is seen during or after a tissue-damaging procedure but not when local anaesthesia is used; (3) identify pain, if it is present after a tissue-damaging injury but is not present when effective analgesics are used; and (4) be injury-specific.

The first axiom is intuitive. The posture, demeanour, activity and vocalisation of injured animals, which differ from those of their healthy peers, are used routinely as indicators of pain. Thus, it is possible to identify specific behaviours that occur only, or more or less frequently, after specific injuries if the injured animals and their uninjured peers are observed and their behaviours compared. This is fundamental to the human-based methods of assessing pain in animals. However, these behaviours may indicate not just pain, but irritation, dysfunction or convalescence.

In the second axiom, if a behaviour occurs during or after treatment but does not occur when local anaesthesia is used, then that behaviour is probably evoked by nociception elicited within or near the damaged tissue.

Different injuries may elicit unique behavioural responses because the sensations experienced by an animal may differ when different tissues are injured or similar tissues are damaged in different ways. The unique behaviour evoked by specific injuries makes it difficult to use behaviour alone to compare the pain experienced by animals subjected to different treatments. If behaviour is to be used to compare the pain experienced by animals, there needs to be a continuum of expression of a single behaviour in response to different treatments (Lester *et al.*, 1996).

Overt behaviour is the final outcome of a complex interaction between physiological responses and psychological influences. The pain caused by a particular injury may stimulate an animal to behave in a pain-reducing manner, but this may conflict with the animal's desire to remain inconspicuous (Broom, 2000). Overt behaviour may not be the best index of pain because of these complex influences on it, but with careful observation behaviours indicative of pain or irritation may be observed hours, days and weeks after injury. In dogs, the effect of analgesia on these behaviours has not been studied enough to confirm that they are definitively pain-related, and furthermore it may not be possible to differentiate between pain and irritation.

Fox *et al.* (2000) identified 166 behaviours in bitches, subject to ovariohysterectomy, that could be used an indices of post-operative pain; 76 of these occurred so infrequently as to be of no clinical use. Five behaviours frequently seen associated with surgery were a decrease in cage circling speed, an increase in drawing up the hindlimbs, incision licking, vomiting, and flank gazing.

Hardie *et al.* (1997) identified 15 behaviours and eight combinations of wakefulness and body position that were useful to characterise post-operative pain behaviour. Hansen (2003) described a programme in which the behaviour of dogs was automatically measured for 24 hours after anaesthesia, anaesthesia and ovariohysterectomy, or partial enterotomy. The dogs had been given a placebo or the analgesic carprofen. Trained caretakers evaluated the pain using a visual analogue scale (VAS), a four-level verbal-rating scale, and a categorised numerical-rating scale (NRS). Some behaviours (time spent near front of cage, speed of movement, distance travelled), plus the pain-rating scales, could identify a 'surgery effect' but not a 'carprofen effect'. This study illustrates the weakness of using overt behaviour *per se* and scales, and suggests the importance of combining behavioural and physiological responses to painful stimuli.

Pain threshold studies, using toe-pinch callipers (Hamlin *et al.*, 1988) or a hand-held pin-pushing device (Lascelles *et al.*, 1997; Slingsby *et al.*, 2001), have been used on dogs to examine the analgesic effects of a number of drugs. The threshold to a mechanical stimulus changed in animals suffering from chronic pain (Ley *et al.*, 1989), which might be due to changes in nerve function or nociceptive processing at a higher level (Nolan, 2000b). Sheep with severe lameness had significantly lower thresholds to a mechanical stimulus than sound animals, and this hyperalgesia remained for several months. The threshold for sheep with mild footrot was not different from that for sound sheep (Ley *et al.*, 1995).

When given a choice, animals will probably select the less painful of two experiences or may self-administer analgesics. The latter method has been used to determine the painfulness of lameness in broiler chickens (Danbury *et al.*, 2000) and laboratory animals (Flecknell, 2000), but it has not been used in dogs.

The physiological mechanisms of pain, including the stimulation of nociceptors, transmission along pain pathways, and electrical and neurotransmission activity in the somatosensory cortex, can be monitored (Mellor *et al.*, 2000). The physiological stress responses that are induced by activity in the pain apparatus can also be monitored (Mellor *et al.*, 2000). These stress responses, particularly the plasma cortisol response, have been used frequently in studies designed to assess the pain and distress caused by surgery in livestock (Mellor & Stafford, 1999) but have been rarely used in dogs (Fox *et al.*, 1994; Hansen *et al.*, 1997). Kyles *et al.* (1998) found no difference in respiratory and heart rates, and arterial blood pressure in bitches in the hours following an ovariohysterectomy with analgesia and those that had analgesia alone. Heart and respiratory rates and pupil dilation were poor indicators of pain when compared with a NRS in dogs following surgery (Holton *et al.*, 1998b). However, the value of the NRS used as an indicator of pain also remains unproven, being a subjective scale based on the experience and feelings of the persons involved. To confirm that the physiological parameters are ineffectual would require a study in which there were sufficient control groups, including groups not receiving analgesia (Table 1).

Activity of the sympathetic adrenomedullary system concerned with 'fight-flight' responses may be assessed using plasma adrenaline and noradrenaline concentrations, heart rate and blood pressure. The hypothalamic-pituitary-adrenocortical (HPA) system initiates long-lasting metabolic and anti-inflammatory responses.   Indices of HPA activity include plasma concentrations of cortisol, adrenocorticotropic hormone (ACTH) and corticotropin releasing factor (CRF).   These are useful indices because, within certain limits, HPA activity increases in a graded way in response to the presumed noxiousness of different experiences (Mellor & Stafford, 2000). In clinical research, the plasma cortisol response has been shown to be a useful indicator of pain and distress in dogs (Fox *et al.*, 1994).

Changes in these parameters provide an indication of how unpleasant an experience is emotionally and physically.   The rapid response of the sympathetic-adrenomedullary system makes it useful during the first few minutes after application of a painful stimulus, and the HPA axis is more useful subsequently. The HPA system responds to a wide range of physical and emotional experiences, both unpleasant and pleasant. If the stimuli are obviously noxious and appropriate control groups are used, the plasma cortisol response can be informative about the relative pain experienced by similar animals treated differently, provided sufficient control groups are included in the protocol.

The plasma cortisol response can vary in complexity.   It may be simple, that is rising to a peak and then returning to pre-treatment concentrations, as is usual with castration and/or tailing of lambs (Dinniss *et al.*, 1997), or it may be more complex, as with amputation dehorning of calves (Sylvester *et al.*, 1998a,b).   A plasma cortisol concentration-time curve derived from repeat blood samples allows the magnitude and speed of change, and the duration and pattern of the entire response, to be determined.   Differences between groups in the initial or later concentration changes, peak concentration and time to reach it, and time of return to pre-treatment concentrations are informative, but only if they can be related to the entire response.   In contrast, comparing plasma cortisol concentrations before treatment with values at one or two arbitrary times afterwards provides little valuable information and can be misleading. The plasma cortisol response can be quantified in several ways (peak height, response duration, area under the cortisol curve), but no single numerical factor adequately defines the response, and the more complex a response, the less likely it is that one number will represent it effectively.

Thus, the plasma cortisol response has been used to assess pain but care needs to be taken when interpreting responses at the upper extremes of the response range, because different noxious stimuli applied simultaneously may not have additive effects on cortisol responses. This is a 'ceiling effect', in which the overall noxiousness of two painful stimuli applied simultaneously may be underestimated as each alone would elicit a maximum cortisol response (Molony & Kent, 1997; Molony et al., 2002; Wilson & Stafford, 2002).

A wide variation in cortisol responses to a specific treatment is common. It is important to distinguish between the effects of pre-treatment stressors on different animals and animal-specific differences. Some animals show consistently high and others consistently low responses to the same stimulus (Petrie et al., 1996a, b). The use of cortisol as a single physiological index allows conclusions to be drawn only about those features of the acute response that are reflected by changes in plasma cortisol concentrations (Mellor & Stafford, 1999, 2000). The return of plasma cortisol response to pre-treatment concentrations does not prove that the animal is pain-free; it merely indicates that the noxious input at that time is not sufficient to elevate cortisol.

The plasma cortisol response in bitches following ovariohysterectomy was sustained which suggests that the physiological and psychogenic inputs are important for between 5 and 24 hours (Fox et al., 1994). The integrated plasma cortisol response for 6.5 hours following treatment increased in the following order: control, anaesthesia, analgesia, and ovariohysterectomy. When an analgesic (oxymorphone) was given to bitches undergoing ovariohysterectomy there was a slight decrease in the plasma cortisol response (Hansen et al., 1997). When Fox et al. (1998) gave butorphanol after ovariohysterectomy it reduced the plasma cortisol response significantly, although when it was given pre-operatively it did not.

3.2   Human-based methods

The difficulties in putting together the scientific protocols for the animal based methods described above encourage many to use anthropocentric techniques to evaluate pain in dogs. These researchers look at the overt behaviour of the dog and, with knowledge of what the dog had experienced, they quantify that pain, using one of several systems. It is accepted that the behaviour and demeanour of an animal are indicative of pain, but when an injury or a disease incapacitates an animal physically and its normal functions are not easily performed, then the behaviour may reflect the degree of incapacitation rather than pain. Again, if specific behaviours are important for convalescence (Hart, 1991), they may be part of the recovery process rather than be caused by pain. Incapacity may cause distress by limiting an animal's ability to escape from predators, but this is not pain. Behaviour following injury may increase an animal's chances of survival by preventing further damage, maximising wound healing, stopping another animal from inflicting more damage, or eliciting help from con-specifics (Molony & Kent, 1997), but may not occur in response to pain.

Table 2. A pain assessment sheet for use in dogs.

Assessment	Pain exhibited
Observe from outside cage:	
Posture	Rigid
	Hunched or tense
	Neither of these
Does the dog seem to be	Restless
	Comfortable
If the dog is vocalising is it	Crying or whimpering
	Groaning
	Screaming
	Not vocalising/none of these
If the dog is paying attention to its wound is it	Chewing
	Licking/looking/rubbing
	Ignoring its wound
Open cage, call dog to come:	
Does the dog seem to be	Aggressive
	Depressed
	Disinterested
	Nervous/anxious/fearful
	Quiet or indifferent
	Happy and content
	Happy and bouncy
Get the dog to stand and sit or walk:	Stiff
During this procedure did the dog seem	Slow/reluctant to rise/sit
	Lame
	None of these
	Assessment not performed
When the wound was touched did the dog	Flinch
	Snap
	Growl or guard wound
	None of these

Adapted from Holton *et al.* (2001)

NRS and VAS were unreliable at assessing acute pain in dogs in a hospital setting (Holton *et al.*, 1998a), and there was a poor correlation between a NRS and the physiological parameters associated with pain in dogs (Conzemius *et al.*, 1997; Holton *et al.*, 1998b). Lists of behaviours which indicate that a dog is in pain have been produced (Morton & Griffiths, 1985; Sandford *et al.*, 1986; Dodman *et al.*, 1992), but these have not been validated. Other scales to assess pain in dogs have been developed (Conzemius *et al.*, 1997; Hellyer & Gaynor, 1998; Firth & Haldane, 1999) but these were poorly defined. However, Holton *et al.* (2001) has defined words used to describe pain in dogs and developed a scale to measure this. He classified a set of 47 words, utilised by veterinarians to describe pain, into nine categories (demeanour, response to people, response to food, posture, mobility, activity, response to touch, attention to painful area, vocalisation), plus a physiological category. These terms were further evaluated by veterinarians and a score sheet produced to be used in the clinic to evaluate pain in dogs (Table 2). Mathews *et al.* (1996) compared the effects of three different analgesics using a subjective scale with four ranks of behaviour (exuberant, normal, comfortable, and assumed painful) and rated vocalisation, movement and respiratory patterns on a scale of three. The assumed painful behaviours included crying, depressed, shaking, thrashing, biting, reluctance to move, and splinted abdomen.

However, this score sheet has not been used in a clinical setting to see if it is differentiate between animals receiving analgesia or not, and it has not been compared with the physiological responses of dogs to painful stimuli.

3.3   Veterinarians in practice

In one Australian study Watson *et al.* (1996) found that most veterinarians used vocalisation, response to handling, and depression as indicators of pain in dogs (Table 3).

Table 3. Signs used by 471 veterinarians as indicators of pain in dogs.

Indicator of pain	No. of veterinarians noting signs
Vocalisation	271
Response to handling or palpation	193
Depression	135
Other behavioural changes	129
Altered respiration	109
Immobility	104
Abnormal posture or function	95
Inappetence or anorexia	90
Increased heart rate	83
Muscular movements	59
Aggression	52
Restlessness	44

Adapted from Watson *et al.* (1996)

The behaviours listed in Tables 2 and 3, and those used in a more recent study by Muir *et al.* (2004) (vocalisation, movement and interactive behaviours), differ from those identified by Fox *et al.* (2000) as being closely related to the pain caused by ovariohysterectomy. It is known that the pain caused by different types of surgery, or disease of different tissues cause different behavioural responses and so it is necessary to identify the behaviours relevant to major types of surgery and use them as indicators of pain, rather than using a range of intuitive but perhaps irrelevant behaviours. In rats, Roughan and Flecknell (2001) identified four of more than 100 behaviours seen after surgery that were useful as clinical indicators of pain. This is similar to the number of behaviours identified by Fox *et al.* (2000) as being useful indicators of pain in bitches after ovariohysterectomy. Thus, it is necessary to identify the relevant behaviour in a series of behavioural studies in dogs if we are to ensure that specific analgesics are working in the short and longer term.

The option of monitoring a mixture of behaviours and physiological responses (Laredo *et al.*, 2004) and various scales (Hudson *et al.*, 2004), by veterinary staff or owners (Vaisanen *et al.*, 2004), to determine if a dog is experiencing pain remains more popular than identifying a limited number of clinically useful behaviours determined by specific research protocols.

Defining the relevant behaviours is ethically costly and financially expensive and tedious. An alternative is to accept that dogs experience pain after surgery, and using the awareness of certain conditions known to cause pain in humans, administer analgesics without monitoring for pain.

4.   ANALGESICS

The identification of morphine as an analgesic was a major breakthrough in human medicine, but until recently interest in the alleviation of pain in dogs has concentrated on making surgery easier and safer by using general anaesthesia or deep sedation. Post-operative analgesia was not considered. Surgery results in acute pain, which is experienced during the procedure and for some hours or days afterwards, and in the development of hyperalgesia where the response to a painful stimulus is exaggerated, or allodynia when stimuli that are normally not painful become painful. Pain management should reduce acute pain and prevent pathological pain (Nolan, 2000a).

General anaesthesia acts to prevent the perception of pain during surgery, but pain may be perceived immediately after the animal becomes conscious again. Therefore anaesthesia should be combined with a longer-acting systemic analgesic to prevent post-operative pain. Local anaesthesia is used to prevent pain during a surgical procedure but is not commonly used in dogs. Lignocaine, which is the common local anaesthetic drug used in veterinary practice, is applied to nerves or the spinal cord and is metabolised very quickly once absorbed systemically. It is short-acting, lasting for about 45 minutes. Local anaesthetics have the disadvantage that they block motor as well as sensory nerves, which can cause temporary paralysis.

There is now a range of systemic analgesics, mostly either opioids or non-steroidal anti-inflammatory drugs (NSAIDs) (Table 4), and there are many clinical indications for their use in dogs (Flecknell and Waterman-Pearson, 2000). The analgesic effect of NSAIDs was initially considered to be due to their peripheral anti-inflammatory actions, but some of them have also been shown to have central nervous system effects (Chambers et al., 1995). The list of neuromodulators with which NSAIDs interact directly or by altering transcription is growing daily. The effectiveness of any particular drug as an analgesic will depend on its pharmacokinetics, on the cause of pain, and on poorly understood characteristics of the drug itself. Some NSAIDs, notably phenylbutazone, are more anti-inflammatory than analgesic; while others, such as carprofen, are more analgesic than anti-inflammatory. The longer period of activity of some NSAIDs may make them more useful in clinical practice than opioids. Pharmacokinetics varies widely between species and extrapolation between them is problematic.

Morphine is very effective for the control of pain in dogs. Pethidine is a useful analgesic if given intramuscularly, and methadone, oxymorphone and buprenorphine are also useful analgesics (Dobromylskyj et al., 2000). Veterinarians in Australia commonly use the narcotics pethidine and buprenorphine and the NSAIDs flunixin and dipyrone in dogs (Table 4). Some NSAIDs, such as phenylbutazone, have a very long half-life, but aspirin has a very short one. Pethidine has a very short duration of activity (Table 4) and its widespread use in Australia surprised Watson et al. (1996). Some drugs are better at treating different types of pain than others. In humans, morphine is effective against inflammatory pain but not neuropathic pain, while tricyclic antidepressants are more useful against neuropathic pain than inflammatory pain (Muir & Woolf, 2001). There is little information about the use of the latter as analgesics in dogs.

In the last decade, there have been many trials investigating the comparative analgesic effects of different drugs in dogs, both when used pre-inter- and post-operatively (Branson et al., 1993; Nolan & Reid, 1993; Lascelles et al., 1994; Pibarot et al., 1997; Grisneaux et al., 1999; Slingby & Waterman-Pearson, 2001; Caulkett et al., 2003; Fowler et al., 2003). This body of research, and its publication at veterinary conferences and in a wide range of journals read by veterinarians in clinical practice, is stimulating and supports the increased use of analgesics by veterinarians, as does the large number of reviews of pain assessment and its alleviation in dogs (Potthoff & Carithers, 1989; Johnson, 1991; Hanssen, 1994; Brock, 1995; Abeynayake, 1997; Fox, 1999; Hellyer, 1999b, 2002).

Table 4. Drugs used as analgesics in dogs by 481 veterinarians (vets) in Australia.

Drugs	No. Vets using drugs	Potency*	Dose (mg/kg)	Intervals (hrs)
Narcotic analgesics:				
Buprenorphine	36	25	0.01–0.02	12
Butorphanol	11	5	0.4–0.6	4
Fentanyl	7	100	0.04–0.08	2
Methadone	8	1	0.2–1.0	2–6
Morphine	14	1	0.2–1.0	2–6
Pethidine	63	0.2	2–10	2
NSAIDs:				
Aspirin	25			
Copper indomethacine	10			
Dipyrone	43			
Flunixin	49			
Ketoprofen	22			
Paracetamol	1			
Phenylbutazone	29			
Piroxicam	1			
Alpha-2 agonists:				
Xylazine	34			

* Relative to morphine
Adapted from Watson *et al.* (1996)

5.   VETERINARIANS AND ANALGESICS

Until recently, the use of systemic analgesics in companion animal practice has been low, but within the last twenty years veterinarians have begun to use analgesics more commonly, especially after surgery thought to be particularly painful. In Britain and in Canada, more than 93% and 84% of veterinarians, respectively, used analgesics with orthopaedic surgery (Dohoo and Dohoo, 1996; Capner *et al.*, 1999), but only about 50% of dogs received analgesia after ovariohysterectomy (Capner *et al.*, 1999). Similarly, more than 80% of veterinary respondents in an Australian study used analgesics for what they considered painful surgical procedures and 94% used them for acute severe trauma (Watson *et al.*, 1996). Veterinary use of analgesics appeared to be influenced by assessment of the painfulness of the procedure, and they were used by 68% for thoracotomy, 60% for cruciate ligament surgery, 53% for lateral ear resection, 34% for mastectomy, 32% for dentistry, 29% for perineal hernia repair, and 22% for toe amputation, but only 6% for ovariohysterectomy and 4% for castration (Watson *et al.*, 1996). In the late 1980s, only 40% of dogs undergoing major surgery received analgesia in one North American veterinary teaching hospital (Hansen & Hardie, 1993). In 1999, more than 80% of veterinarians in South Africa did not intentionally use analgesics in dogs undergoing sterilisation (Joubert,

2001), and in the mid-1990s in Canada, only 13% of veterinarians used analgesia following ovariohysterectomy and 11% following castration (Dohoo and Dohoo, 1996). A recent survey in New Zealand found that 76% and 65% of veterinarians used analgesics for dogs undergoing ovariohysterectomy and castration, respectively (Williamson *et al.*, 2005).

Veterinarians tended to give analgesics during or after surgery but before anaesthetic recovery, with few waiting until recovery to see if pain developed (Watson *et al.*, 1996). The former is the usual approach in human surgery also. There is some support for giving analgesics before surgery commences, as this prevents the occurrence of painful stimuli and patients may require less post-operative analgesia. The acute pain caused by surgery causes an increased stress response (Fox *et al.*, 1994) which prolongs recovery and increases morbidity (Lascelles *et al.*, 1994). There may also be prolonged hyperexcitability of neurons in the spinal cord, leading to tenderness, hyperalgesia, bouts of intense pain and reduced efficacy of analgesics (Lascelles *et al.*, 1994), thus supporting the use of analgesics pre-operatively. Moreover, leaving analgesia until after surgery is the least efficient way of using analgesics. The pain experienced by dogs following surgery has been predicted from the human experience of different types of surgery (Table 5). However, there have been no attempts to compare the physiological and behavioural responses to different types of surgery in dogs.

Table 5. Expected post-surgical pain in dogs.

Surgery	Expected level of pain
Nail trim, suture removal, cast application	No pain
Suturing, debridement, urinary catheterisation, dental cleaning, ear examination and cleaning, abscess lancing, cutaneous foreign body removal	Minor pain
Ovariohysterectomy, castration, Caesarian section, cystotomy, anal sacculectomy, dental extraction, cutaneous mass removal, laceration repair	Moderate pain
Fracture repair, cruciate ligament repair, thoractotomy, laminectomy, amputation, ear canal ablation, exploratory laparotomy, ophthalmological surgery	Severe pain

Adapted from Johnson (1991) and Hellyer (1999).

In Canada, the presence of an animal health technician in a veterinary practice modified attitudes towards pain and increased the likelihood of analgesia being used, possibly due to the technican being involved in observation of the animals and their treatment in the wards (Dohoo and Dohoo, 1996). Female veterinarians had a similar effect (Dohoo and Dohoo, 1996; Capner *et al.*, 1999). Veterinarians who had graduated within 10 years were less concerned about the side-effects of analgesics, and used analgesics more commonly than those who graduated earlier. Of concern was the observation by Dohoo and Dohoo (1996) that attendance at continuing education activities within the previous 12 months increased veterinary concern about the side-effects of opioid analgesics and reduced the usage of analgesics.

The administration of analgesics to dogs post-operatively does not mean that they receive effective pain relief, as there is little information concerning effective dosage rates and the duration of effectiveness for most analgesics (Capner et al., 1999). In addition, administering one analgesic is unlikely to be 100% effective, and inflammatory pain, neuropathic pain and cancer pain are best treated by a combination of analgesics that act by different mechanisms and possibly synergistically (Muir & Woolf, 2001). When given peri-operatively, analgesics are usually continued for a maximum of 24–36 hours. It is likely that after some orthopaedic surgery a dog may experience pain for longer than that, but little is actually known about the effect of analgesia on the longer-term behaviour of dogs after surgery. Long-lasting-depot-analgesics make the long-term use of analgesics more feasible. The use of analgesics in research laboratories, including for dogs where invasive surgery is practised, is now almost universally mandatory, but, surprisingly, they are not always used (see Chapter 9).

Chronic pain is usually defined as pain present for 3–6 months, but veterinarians usually consider a dog to be in chronic pain if it has persisted for 3–4 weeks (Hardie, 1996). Chronic pain may cause significant changes in the nervous system, including hyperalgesia and allodynia. Common causes of chronic pain in the dog include osteoarthritis, degenerative disc disease and cancer. Chronic pain is difficult to identify if the behaviour of the dog is not altered, and may be difficult and expensive to treat. Pathological pain in the neck and back occurs in many medical conditions in dogs and may be difficult to diagnose (Webb, 2003).

Osteoarthritis is the most common joint disease in dogs and in the UK, owners of at least 500,000 dogs out of a population of more than 6 million dogs have sought treatment for this disease; it is estimated 1.3 million dogs probably suffer from it. In the USA, probably 10 million dogs suffer from the condition at any one time and only a small number of these will be treated (Lascelles & Main, 2002). In one study of military dogs, 20% of those humanely killed had osteoarthritis (Moore et al., 2001). Veterinarians use analgesics to treat the pain associated with osteoarthritis, in combination with chondroprotective agents, weight reduction and controlled exercise (Hardie, 1996), but many of those treated will not receive sufficient analgesic or for an insufficient duration.

As long ago as 1978, Yoxall recommended analgesia for dogs with cancer, but because the signs of pain are not as obvious as in osteoarthritis, owners and veterinarians may be slow to alleviate it. Lascelles & Main (2002) estimated that in the USA, with a population of more than 50 million dogs, 20–25% (8–12.5 million) might have cancer. The level of pain caused by different cancers will vary from none to extreme pain, but this may not be recognised in many of these animals. If recognised, treatment is available but how beneficial it is remains unclear. Cancer pain may need to be treated stepwise, using a NSAID initially combined with a weak and then a strong opioid analgesic. If nerve compression due to a nerve root tumour is causing pain then it may be necessary to use antidepressant drugs such as amitriptyline (Hardie, 1996).

The reluctance to use analgesics, even in companion animals, is difficult to understand, given that almost all veterinarians believe that animals feel pain. It is curious that many clinicians will use antibiotics without being certain that a bacterium sensitive to that antibiotic is involved in the disease, yet they may be loath to use analgesics unless certain that the animal is in pain. However, veterinarians are concerned about the side-effects of narcotic and NSAID analgesics (Capner et al., 1999), but there is now a wide range of analgesics with few side-effects to allay these concerns. In addition, many veterinarians may be reluctant to use opioid analgesics because of the problems with safe storage and the associated paperwork (Watson et al., 1996; Capner et al., 1999), but cost does not seem to be an issue restricting the use of analgesics in dogs (Capner et al., 1999).

There is an argument against using analgesics, which states that pain causes inactivity and this promotes healing and accelerates convalescence; 30% of veterinarians surveyed thought that pain could limit a dog's activity after surgery and so reduce risk of suture breakdown (Capner et al., 1999). Ther is little evidence to support this concern. From a professional standpoint, veterinarians should be advocating the use of analgesics, wherever possible, as they are ethically obliged to prevent unnecessary and unreasonable pain. Younger veterinary graduates and women in particular award higher pain scores for ovariohysterectomy than older graduates and men, respectively (Capner et al., 1999). This suggests that increasingly, veterinarians will use analgesics routinely in future for almost all surgery.

In a useful review, Pascoe (2002) discussed the use of alternative that is non-pharmacological, methods of alleviating pain in animals. There is little published in this field, but acupuncture may relieve pain in humans. However, gold wire or gold beads placed in predetermined acupuncture points had no beneficial effect on dogs with hip dysplasia (Pascoe, 2002).

6.   CONCLUSIONS

The prevention and alleviation of pain is a key component in the management of dogs, and is one of the most important ways to improve the welfare of dogs undergoing surgery or in the treatment of chronic conditions such as lameness. There are difficulties with the assessment of pain due to its subjective nature, but until knowledge allows surety about the experience of pain in any animal it is probably best for veterinarians to afford the benefit of the doubt and use analgesics. However, there is a need to further the work of Fox *et al.* (2000) and use defined protocols with sufficient control groups to determine whether what we believe are good behaviours to monitor pain in dogs are actually useful. Chronic pain may be a much greater problem in dogs than acute pain. The diagnosis of chronic pain is only possible if the animal's behaviour changes significantly. This may not occur with many conditions, and even if recognised treatment may be inadequate.

Chapter 7

TRAINING METHODS

Abstract: There are more books about the training of dogs than the training of any other
              species. This suggests that there is great interest in the subject and that there
              are many trainers who believe their advice is valuable. Most modern textbooks
              relate to the dog as a companion rather than a working animal. Indeed, there
              are few books dealing with the training of dogs that work with sheep or cattle.
              Most books recommend positive reinforcement as the most important training
              technique. Inadequate training and a lack of control are a major causes of dogs
              been abandoned, neglected and of public disquiet about canine behaviour. The
              misuse of punishment and training tools such as electric training collars has a
              negative effect on the welfare of dogs and may result in fearful, timid dogs
              that are stressed even by the presence of their owner or handler. However,
              there is not enough known about the physiological responses to different
              methods of training and whether training dogs for sports such as agility is
              stressful.

1.   INTRODUCTION

The methods used to train and retrain dogs are described in a multitude of books, suggesting there is a cadre of dog-orientated people interested in these topics. This may include a large number of people who want to know how to train their dog but do not want to attend classes, or have attended classes and not found them useful, or who have dogs with behavioural problems. In addition, books are available on training dogs how to engage in particular sports or work.

The desired outcomes from training a companion dog are relatively few and straightforward. The animal should always come when called, sit and stay, lie quietly, understand "NO!" and "be quiet" and should walk beside

129

the handler and generally ignore other people and dogs. The dog should not be aggressive and be able to cope with being alone for long periods of time. It should be trained to toilet outside, preferably in one place on the owner's property, or indoors in a toilet tray. It should not jump up on people nor bark incessantly.

Working and sporting dogs are expected to accomplish whatever they were bred, reared and trained to do effectively and safely. If competitive, they should win or at least perform well. Show, agility, obedience, field trial, and racing dogs have to be trained to behave in a manner which facilitates their success. In competitive obedience dogs have to perform in quite a specific manner, but many obedience classes focus on training dogs to be under control in private and public, that is becoming good canine 'citizens'.

Dogs were probably selected for an ability to focus on human behaviour and are very sensitive to human body language such as pointing (Miklosi et al., 1998; Hare & Tomasello, 1999; Soproni et al., 2001, 2002). They will respond to humans pointing even when olfactory cues suggest that food is not placed where the person is pointing (Szetei et al., 2003). One dog, a 9-year-old Border Collie named Rico, was shown to be very sensitive to words and learn to associate over 200 items with spoken words. It had an amazing ability to learn what new words meant in a single trial (Bloom, 2004; Kaminski et al., 2004). If dogs are so focussed on humans then many of the difficulties seen with uncontrolled dogs are probably due to humans giving the wrong signals and inadvertently encouraging inappropriate behaviour. Thus the difficulty in training dogs to behave in a quiet and controlled manner is probably due to humans not reinforcing such behaviour.

Dogs, regardless of breed, are very attached to humans and regard them as a secure base (Topal et al., 1998). This should make the training of dogs to behave appropriately easy, which it is, but many people do not know the basics of training. In addition some people in the dog training industry may want to make dog training sound difficult and complicated to enhance their role. Training in dog schools may not be appropriate for some dogs and owners, and training at home may be more effective. Dogs have to be taught what is correct as well as what is not acceptable. As dogs are very sensitive to our actions it is surprising there are so many dogs with behavioural problems, and this suggests the physical and social environment which many dogs inhabit is inadequate, that is too little to do and too little contact with humans.

In essence, virtually all training is undertaken using different aspects of operant and classical conditioning, to suit the animal, owner and behaviour required. Books describe different techniques but the theories underlying them are all essentially similar. There are different schools of thought about dog training; one suggests that dogs should be trained with positive reinforcement only, and another that dogs should made 'respect' the handler without any overt reinforcement. Historically, punishment and negative reinforcement were the main methods used to train dogs but now positive reinforcement is more popular (Hiby *et al.*, 2004). Methods vary from friendly methods, such as 'clicker training', to quite harsh methods, which depend on positive punishment including beating and the use of electrical shock collars, punishers that could wound and cause mental harm (Schilder and van der Borg, 2004). Generally, pet dogs are trained using positive reinforcement while working dogs are trained using positive reinforcement and punishment. In this chapter, the welfare implications of training methods and poor training will be discussed.

2.   INADEQUATE TRAINING

The most important aspect of training with regard to the welfare of a dog is the effect often inadequate training. Evidence of poor training includes uncontrolled aggression, jumping up on people, hyperactivity, and excessive barking. In a survey of dog owners in suburban Melbourne, the majority (63%) of respondents stated that their dog was overexcited, jumped on people (56%), rushed people (38%), or barked excessively (32%) (Kobelt *et al.*, 2003b). The majority of these dogs had received some training from the owner but 35% had no training at all and only 20% had formal training. This was similar to the 24% of dogs, in an American study that had formal training (Voith *et al.*, 1992). Large dogs were more likely to have had formal training (Kobelt *et al.*, 2003b). The type of training is important with regard to the welfare of the dog during training, but the outcome is very important as it impacts on the dog's whole life experience.

If formal training is likely to result in more obedient and controlled dogs then it is to be recommended. There is evidence that attendance at formal obedience classes does not influence the occurrence of problem behaviours (Voith *et al.*, 1992; Voith, 1993), but this was not found in other studies by Clark and Boyer (1993) and Jagoe and Serpell (1996). Puppy socialisation classes appear to have short-term effects on the obedience in dogs (Seksel *et al.*, 1999). Obedience training undertaken by only a small proportion of dog owners caused an improvement in relations between dog and owner (Clark & Boyer, 1993). Many working and sporting dogs are killed despite being well bred and having the ability to be capable workers, because they are badly trained and do not reach their potential. These dogs are often neglected and abused before finally being disposed of.

Dogs which are better trained are less likely to have behavioural problems. Beaver's (1994) analysis of surveys from veterinary clients and the general dog owning public showed that aggression was the most common behavioural problem being found in 19% of dogs. However, excessive barking (13%), begging for food (8%), jumping on furniture (6%) or people (6%) were also significant problems, suggesting a widespread problem with disobedience. Disobedient dogs are less likely to be taken for walks or played with and are probably more likely to be neglected, abused, killed or surrendered to a shelter. Poor or inadequate training and its behavioural consequences are likely to impact seriously on the welfare of individual dogs. Well-behaved dogs are more likely to live indoors, be treated as family, be valued, and brought to the veterinarian and groomer, than disobedient dogs, or those out of control. Owners are less likely to spend time and money on dogs that defaecate indoors, jump up on visitors and attack the neighbours.

3.   TRAINING TECHNIQUES

Learning theory forms the basis of training techniques. Psychologists have categorised learning in different ways but a common classification includes imprinting, habituation, operant conditioning, classical conditioning, latent learning and insight learning (Drickamer and Vessey, 1982). All these classes of learning are important for our understanding of how a dog learns to cope with its environment and in allowing the dog's owner to train it as required. The key component of training is using the different elements of operant and classical conditioning, but it is important to develop a training programme, be consistent, and have short happy lessons which end favourably. Fear is contraindicated during training.

3.1   Operant conditioning

Operant conditioning is based on the premise that the consequences of a behaviour will determine the probability that the behaviour will or will not be repeated. If the consequence is pleasant (reinforced) then the behaviour will probably be repeated, but if the consequence in unpleasant (punished) then the behaviour may not be repeated. There are four major elements to operant conditioning; positive reinforcement, negative reinforcement, positive punishment, and negative punishment.

Positive reinforcement occurs when a specific behaviour is followed by a pleasant experience, a reinforcer, and thus makes the behaviour more likely to be repeated. This is the basis of all modern dog training. Reinforcers may be primary such as food, play, petting, a toy or sex, or secondary such as praise or a 'clicker'. The former are of biological significance and the latter are associated with primary reinforcers by classical conditioning. Positive reinforcers must be delivered immediately the behaviour occurs. A delay, of even a few seconds, between the behaviour and the reinforcer reduces the efficacy of the latter. During early training, the ratio of behaviour-to-reinforcer should be 1:1, but once the relationship is learned the reinforcer should be given occasionally. Food works best when the dog is a little hungry but not ravenous. Some dogs do not find food to be a very good reinforcer so a plaything might be used instead.

Negative reinforcement involves removing something unpleasant from the environment to increase the likelilhood of the desired behaviour. Choke chains are often used incorrectly as negative reinforcers, as are electric training collars.

Positive punishment is when an undesirable behaviour is associated with an unpleasant experience, a punisher, which must be appropriate and delivered immediately. A punisher is targeted at the behaviour not the animal. Physical punishment is not generally recommended. Verbal punishment is often effective if delivered effectively. Saying NO in a loud voice to a young pup may be an effective punisher. Punishers may be too strong or too weak. The latter result in habituation, the former constitutes abuse. Excessive reliance on punishment may result in the dog becoming fearful of the handler or owner. Punishers are often delivered too late and the behaviour being punished is not the target one. For example, if a dog does not come when called and is punished when it does come then returning to the owner is being punished.

Remote punishment is often recommended for working dogs and electric training collars are used to deliver a remote punisher. The hand should not be used to punish a dog, as the dog may become hand shy, and punishers should not make the dog fearful (Landsberg *et al.*, 1997). Punishment does not mean physical abuse. A loud noise, a growl or a yell, may be sufficient to punish a particular behaviour. Distraction, for example throwing a tin with stones in it or a chain near to the dog is often recommended as a punisher. Poor use of punishment has a significant negative impact on the welfare of the young dog. In one study of how owners attempted to stop their dog digging, many of the methods used were punishers (Adams & Grandage, 1989). These included scolding, smacking, spraying with a water hose, putting faeces or a mouse trap in the hole, holding the dog's head in the hole, putting the dog in the rubbish bin or in the hole, or putting pepper on the dog's nose. Few of these methods on their own will stop a dog digging, and they miss the important element of reinforcing an alternative behaviour and providing alternative activities, or a special digging area. The methods used as punishers show how ignorant and desperate many people are when dealing with their dog's behavioural problems.

Negative punishment occurs when an animal that is experiencing something pleasant engages in an unacceptable behaviour and the pleasant experience is ended immediately. For example, if a dog sitting beside the fire in the living room growls at a cat and it is then put outside immediately.

All four elements of operant conditioning may be used in training dogs, but emphasis should be on positive reinforcement for companion animals and punishment should be used only when necessary. The use of punishment appears to be common. In one survey in the UK, 66% of respondents used vocal punishment and 12% physical punishment, while 60% used praise, 51% food, and 11% play as reinforcers; overall 20% used reward-based methods only, 10% used punishment only, and 60% used both (Hiby *et al.*, 2004). In this study, respondents used different methods of training in different situations (Table 1). There was a positive correlation between reward-based methods and dog obedience score. Dogs were more likely to give up something if trained using play as a reinforcer, more likely to walk-at-heel if trained using a reward-based method, and not to chew items if given an alternative (Hiby *et al.*, 2004). These researchers concluded that punishment was not suitable for everyday training because although it could be very effective if used correctly, pet owners may not understand how to use it properly. Lindsay (2000) in his classic texts on dog behaviour and training, considered excessive physical punishment to be inappropriate but suggested that punishment had a positive outcome in that it enforced social boundaries. He felt that problems with punishment arose from a poor understanding of its uses as a behavioural tool.

Clicker training of dogs is now very popular. The clicker is a conditioned secondary reinforcer which has the advantages of being consistent, easier to use than many other reinforcers, and it is not influenced by the emotional status of the trainer as is a voice reinforcer. The clicker can also reinforce a behaviour carried out at a distance from the trainer and is an ideal reinforcer for shaping behaviours through gradual approximations.

Table 1. The percentage of respondents in a UK survey who used different training methods for dogs for specific tasks.

Task	Punishment	Reward	Various
Toilet training	12	39	45
Chewing objects	79	4	40
Stealing food/objects	84	7	10
Sit on command	75	39	
Come on command	2	78	
Leave object	30	63	
Heel training	26	45	

Adapted from Hiby *et al.* (2004)

Dogs trained by harsh methods of punishment are more submissive and likely to engage in paw lifting, snout licking and crouching, all of which may indicate stress or distress (Schwizgebel, 1983, cited by Beerda *et al.*, 1997). Increased stress may have implications for immunocompetence and health. Dogs, whose owners used shaking or hitting during training, were more likely to be aggressive towards other dogs (Roll & Unshelm, 1997) and might show increased anxiety and separation-related behaviours, whereas reward-based training was associated with reduced hyperactivity (Hiby *et al.*, 2004).

During training, some police and army dogs developed lowered levels of immunoglobulin A in saliva, suggesting stress or distress caused by training, while others maintained high levels (Skandakumar *et al.*, 1995). The latter tended to be confident animals closely attached to their handler. Similar responses were seen in heart rates when trainee guide dogs were walked. Some excitable, stress-prone dogs were unable to temper large changes in heart rate while other calmer dogs could (Vincent & Leahy, 1997). Guide dogs are trained using operant conditioning both positive reinforcement and positive punishment (Lloyd, 2004), which produces very useful dogs (Naderi *et al.*, 2001). The trainers are, however, professional, whereas most people who train pet dogs are amateurs. This suggests that for many dogs training is an unpleasant experience.

The suggestion that it is good for dogs to engage in agility and obedience-type classes and competitions may not be true. Such activities may not benefit the animal's welfare but may be distressing during training and competition. Not enough is known about the effects of training and different training methods on the physiological and immunological responses of dogs.

3.2   Classical conditioning

Classical conditioning occurs when an unconditioned stimulus and a conditioned stimulus are paired to produce an unconditioned involuntary response. After repeated pairing the conditioned stimulus evokes a conditioned response. Examples are the gastrocolic reflex which can be used to train pups to defaecate in a particular place, and when secondary reinforcers are linked to primary reinforcers. It also occurs when dogs develop fear responses to particular stimuli, such as a veterinary clinic.

The lack of knowledge of dog owners and some professional dog handlers and trainers with regard to the theory of learning and its application is an important issue with regard to the welfare of dogs. In addition, training is ongoing so a dog's life is a constant learning experience and therefore it is important for the welfare of the dog that owners understand the fundamentals of learning and training in order to maximise the dog's ability to cope effectively with its environment and behave in an appropriate manner. To condemn the use of punishment without understanding its value and to believe that training should only be by positive reinforcement is pointless (Cheetham, 2003) and does not allow the use of the full range of training options.

4.   TRAINING SHEEP, SPORT AND WORKING DOGS

Sheep dogs drive sheep by barking and a bounding body movement (Huntaways, Bearded Collies), or by using their eyes to stare at and move them (Border Collies, Heading Dogs). In one study of training Border Collies, Marschark and Baenninger (2002) suggested that access to sheep was a positive reinforcer and that denying them access to sheep was negative punishment. In training Border Collies, blocking access (negative punishment) was much more frequently used than allowing access (positive reinforcement). They suggested that this form of punishment was appropriate to that type of training, but that physical abuse was not.

Exercise training for sports dogs should match the level of fitness required. Many sports dogs are under-prepared for their expected level of activity. Hunting dogs are particularly at risk, being held in kennels for most of the year and then expected to work all day under tough physical conditions. Correct training protocols for the type of exercise (Table 2) result in dogs being able to function effectively. Poorly-trained dogs are more susceptible to injury and their inability to work effectively may leave them subject to abuse, neglect and being killed.

Table 2. The type of exercise required for different types of canine activity.

Type of exercise	Type of dog
Sprint	Racing greyhound, whippet and terrier
	Coursing gazehound
	Weight pulling
Intermediate	Hunting (birds, rabbits) or pursuit of game (fox, deer, wild pig)
	Field trial
	Tracking
	Livestock work
	Frisbee, agility, jogging with people
	Police
	Guard
	Military
	Customs
	Border
	Drug detection
	Search and Rescue
	Guide dog and assistance dogs
Endurance	Sled dogs

Adapted from Toll & Reynolds (2000)

5.   TRAINING EQUIPMENT

Choke chains, also called check chains, have a mixed reputation amongst dog trainers. Some regard them as an excellent training tool but others regard them as synonomous with abuse (Mugford, 1995), and their misuse has been criticised (Mugford, 1991; Myles, 1991). When used correctly choke chains are effective punishers, but with poor timing they are likely to mistrain the dog and can be injurious to the neck. When used as a choke they are merely a tool of abuse. The proper use of a choke chain is not learned easily and for that reason they are not suited to novice dog owners. Expecting a new dog owner to learn to use a choke chain on a young dog within an obedience class is unrealistic (Myles, 1991).

Head collars of different types have been available since 1984. They allow owners to control the dog's head and thus the body. They have dramatically improved the success rate for retraining large boisterous or aggressive dogs and allow owners to control what were previously uncontrollable dogs (Mugford, 1995). They are probably the greatest device developed in the last few decades which allow owners to gain control of dogs that otherwise might have been killed or surrendered to a shelter. When a head collar is placed for the first time dogs object to its presence with head shaking, pawing and leash biting (Haug *et al.*, 2002) but this resistance usually wears off after a few minutes. Resistance may be due to irritation or to human control. The head and ears of dogs wearing head collars are usually carried lower, suggesting submission, than when wearing a neck collar and the latter are more unruly (Ogburn *et al.*, 1998). The responses of dogs to four different types of head collar were similar and no one collar was better than the other (Haug *et al.*, 2002). Physiological responses (heart rate, respiratory rate, pupillary dilation, adrenocorticotrophic hormone and cortisol levels) associated with wearing a head collar were not different from those seen when a dog was wearing a neck collar (Ogburn *et al.*, 1998). Most dogs rapidly habituate to a head collar and can be more effectively and safely controlled and retrained. Head collars are easier to use than a choke chain and many dog trainers now prefer head collars to choke chains, this is appropriate and probably to the benefit of the dog (Overall, 1997).

Neck collars are the standard means of control for most dogs, but head collars and chest/shoulder harnesses are also popular. Prong and pinch collars are instruments of abuse and suggest poor training and control. Dog training tools that use sound to punish canine behaviour have not been investigated and so the welfare implications of these tools remain unclear but because they work as punishers there will be some welfare impact, and how significant this is needs to be quantified.

Excessive barking is a constant problem with dogs especially for people living in suburban and urban households. The motivations for persistent or irritating barking must be identified before treatment is started and usually one or a combination of adjusted management, behavioural modification and drugs can be used to stop the barking. However, persistent offenders may be killed, surrendered to a shelter or be surgically de-barked. Anti-barking collars and muzzles have been developed to treat barking by punishing it.

Anti-barking collars are used to treat persistent barking dogs. There are three types of collar, all of which act by punishing barking using an electric shock, a citronella spray, or an unpleasant noise. Barking triggers the collar and activates the punisher. All three are effective but the citronella-spray collar decreased barking in more cases than the electric-shock collar (Juarbe-Diaz & Houpt, 1996). The citronella-spray collar was more effective in reducing barking while travelling in a motorcar than barking stimulated by traffic or television (Wells, 2001). It was also better at training a dog not to engage in coprophagy than an unpleasant noise, probably because the citronella spray is more unpleasant (Wells, 2003). It has been suggested that the citronella spray is a 'disruptive stimulus' rather than a punisher but it fits the definition of punisher in that it reduces the likelihood of barking occurring. If the use of an odourless spray is as effective as a citronella spray then this would be better from the animal's perspective as the smell of citronella makes some dogs vomit, and it is possibly effective because it is so unpleasant. Forcing a dog to smell something unpleasant is probably worse than a minor electric shock or an unpleasant noise, given the sensitivity of the dog's scenting processes. Dogs may have between 220 million and two billion olfactory neurons but humans have only 2–5 million of them.

Punishment is most effective if immediately after the behaviour is punished, the animal is taught a more appropriate behaviour and this is reinforced (Overall, 1997). This is not possible with anti-barking collars. Dogs on which anti-barking collars have been used sometimes engage in redirected aggression and may become anxious and fearful. Punishment alone is not the best way to treat barking and is contra-indicated and cruel when treating barking motivated by anxiety, such as separation anxiety, or when the barking is an obsessive behaviour. Attention-seeking barking and territorial barking can also be treated more effectively by alternative methods. Anti-barking collars are not necessary if the motivation for barking is identified and dog's owners are willing to retrain or treat their dog and/or modify its management.

An anti-barking elasticated muzzle that allows dogs to eat and drink but prevents barking has recently been developed (Cronin et al., 2003). It works by making it difficult for a dog to open its mouth wide enough to bark. When the muzzle was placed, dogs became subdued, lowered their tail and exhibited little tail wagging and less activity than dogs without a muzzle (Cronin et al., 2003). However, the salivary cortisol response was the same in dogs with and without anti-barking muzzles, suggesting that wearing an anti-barking muzzle was not a distressing experience (Cronin et al., 2003). The ability of dogs to drink or pant while wearing the collar was not described but might be important in hot climates.

The use of boundary fences, which depend on an electric shock to train dog to stay away from the boundary, have significant welfare implications if they fail to work and dogs escape and are then exposed to a number of dangers such as road traffic accidents, misalliance and theft. In addition dogs do run through these fences and then cannot get back.

5.1   Electric training collars

Electric training collars have been used on working and pet dogs for more than three decades. These training devices consist of a collar that includes a battery and electrodes which generate an electric shock which can be triggered remotely. The shock can be modified on some collars to suit individual dogs while others have a single strength of shock. Some collars sound a beep before the shock is delivered. The duration of the shock varies from 1/1,000 of a second to 30 seconds (Schilder & van der Borg, 2004), and generally a few thousand volts are used. Electric training collars are generally used as a positive punisher but some are used as negative reinforcers. Electric shock caused pronounced salivary cortisol responses in dogs and changes in behaviour, suggesting distress (Beerda *et al.*, 1998).

When used to train police dogs, electric collars had a long-term effect on the behaviour of dogs compared to those that had not experienced the collar. Dogs trained with the collar engaged in more stress-related behaviours and had lower ear posture than those that had not experienced the collar (Schilder & van der Borg, 2004). They were more stressed on the training ground and in a park than control dogs. The dogs trained by electric collar associated their handlers and getting orders with being shocked. The use of electric training collars during training is a painful and stressful experience and dogs learn to associate their trainer with these shocks, even outside a training context. The way the collar was used by police handlers was important. The shock was used when the dog did not obey the 'let go' command, the dog 'heeled' ahead of the handler, bit the 'criminal' at the wrong time, or reacted too late to the 'heel' command (Schilder and van der Borg, 2004). In these cases the collar was associated with the handlers' orders. Punishing the dog for not immediately obeying the handler's instruction is a poor use of the training collar and will certainly have long-term effects. Some police dog handlers that do not use shock collars use prong collars, beatings and kicks, and choke-collar corrections, and these were not as distressing as the use of the electric training collar (Schilder and van der Borg, 2004).

When electric collars are used to train dogs not to attack sheep the behaviour can be punished without an order being given (Christiansen *et al.*, 2001a). The dog will not associate the shock with the handler. Owners reported no effect on the dog's behaviour during the year following shock training. This type of use may be less stressful and damaging (Christiansen *et al.*, 2001a,b), possibly due to the single behaviour being punished and the defined circumstances in which it occurs. Indeed, use of the electric collar to stop sheep attacks was more effective, and might be less damaging to the dog than some of the common alternatives such as allowing sheep to attack young dogs or using taste aversion (Hansen *et al.*, 1997). The use of electric training collars as a remote punisher in training working sheep dogs to stop biting sheep is sometimes necessary. Most well-bred working dogs quickly learn not to bite sheep when punished effectively in this way, and this may result in improved welfare for sheep and the survival of a good working dog. Alternative forms of punishment used by shepherds may be less effective and more brutal.

Electric training collars have been widely condemned by many individuals and organisations (Cheetham, 2003), but their use has also been praised (Alderman, 2003). When used by trained operators for specific behavioural problems, their use in working dogs might be justified.

However, at present in many countries they are probably greatly overused (Overall, 1997) and much of their use is probably ineffective and causes severe distress to the dogs involved.

6.   DOG TRAINERS AND RETRAINERS

Good dog trainers should be able to communicate well with people, be good listeners, identify with the problems owners have with their dog, and be open to new ideas. If a trainer is using archaic, violent and irrational techniques then the quality of life for the dog and its owner is compromised (Myles, 1991). Because dog trainers are not under any form of professional control in most countries quality probably varies between participants. Thus finding a good trainer is important for new dog owners and for those, like veterinarians and veterinary nurses, who advise them. A good dog trainer can be identified by their success rate and methodology. They listen to owners and ensure that clients understand what is going on; they use methods that are easy for people to master quickly and owners are not expected to master the use of choke chains. The clients of good trainers are happy and relaxed. Good trainers are gentle with dogs and most dogs they train show improvement within one training session and enjoy the training process. Dogs are not threatened or forced and practical exercises for household dogs are taught. Dogs are always on their four feet and not swung or hung off the ground. Good trainers do not use rubber hoses, sticks, leashes and hands to hit dogs, nor do they use pinch collars and electric training collars (Myles, 1991).

Many people work as dog behaviourists and are employed by dog owners to help them train and/or retrain their dog, usually to stop the dog engaging in unwanted behaviours. The development of defined educational packages at a degree or diploma level for people wanting to undertake this sort of work, and national regulation of those involved, would improve the quality of training and retraining dogs and thus the welfare of dogs.

7.   CONCLUSIONS

There is an abundance of sound information available on dog training and thus few excuses for dogs not being trained adequately, but the quality of personnel involved in dog training and retraining is not regulated in many countries. Little research has been conducted to investigate the stress experienced by companion or working dogs during training, and thus the relative stress or distress caused by different techniques and their relative efficiency remain unclear. The lack of adequate training results in abuse, neglect, abandonment and death of many dogs, and miserable lives for both owners and dogs.

Chapter 8

THE WELFARE OF THE ATHLETES; GREYHOUNDS AND SLED DOGS

Abstract: Greyhounds are superb canine athletes. They have few inherited diseases, a beautiful physique and a gentle disposition. However, greyhound racing is a physically demanding sport. There is a disparity between the number of greyhound puppies born and those that make it to the race track. Many are killed because they are inadequate athletes. Dogs are frequently injured and their racing careers are often shortened by injury. Track surfaces and topography have been adjusted to increase speed and reduce injury. Racing dogs are valuable and are generally well managed but when they are finished racing the majority are killed and only a few are kept for breeding. Greyhound adoption clubs attempt to re-home dogs after they have finished racing. Sled dogs race in tough physical conditions and many race in endurance-type races. These dogs have to be well managed to complete these races. Sled dogs are examined by veterinarians before, during and after racing and may be stopped from racing if they are injured, sick or seriously stressed.

1.   INTRODUCTION

The racing greyhound is a superb athlete and greyhound racing is a physically demanding sport. Greyhounds have been selected for the physical and behavioural characteristics required to succeed in racing. Greyhounds can accelerate to speeds of 65 km/hr in a few seconds (Dobson *et al.*, 1988) and at this pace any uncontrolled change in direction or collision with another dog can result in injury. Greyhounds are different from other dogs in that they have higher packed cell volume (PCV), mean cell volume, and haemoglobin concentration, and lower red blood cell (RBC) and platelet counts, lower serum total protein and albumin concentrations (Sullivan *et al.*,

143

1994; Anonymous, 1995). A PCV of 50–65% is normal in greyhounds. The haemoglobin of greyhounds may have a higher affinity for oxygen than that of other breeds (Sullivan *et al.*, 1994). They have lower white blood cell (WBC) counts. The large RBC volume of greyhounds is probably due to a shorter lifespan of these cells in greyhounds than other dogs (Novinger *et al.*, 1996). Greyhounds have a large heart and a slower pulse rate than other breeds. They are particularly sensitive to the barbiturates used in anaesthesia, apparently because of a lower level of oxidative enzymes in the liver (Anonymous, 1995). Heart weight and heart-to-body weight ratios are higher in greyhounds (1.25%) than crossbred dogs (0.80%) (Schneider *et al.*, 1964) and this is associated with racing ability (Schoning *et al.*, 1995).

The health and welfare of the racing greyhound has received a lot of attention. Physiologists have focussed on the physiological changes which occur immediately before, during and after racing. The welfare of greyhounds is of significance to their owners and trainers but also to those who question whether greyhound racing should continue, as it results in serious injuries to many dogs and the killing of dogs after they finish racing. There are other forms of dog racing, including sled racing, and terrier and gazehound racing, but these arc all minor sports compared with greyhound racing. There are about 60 greyhound racing tracks in the USA and in 1990 about 3.5 billion dollars were wagered on dog races (Goodwin & Corral, 1996). In the USA, at each greyhound racing meet there are between 13 and 15 races each with eight dogs. In the USA, it is the 6th largest spectator sport and is legal in 18 states (Noriega & Lin, 2002). Recently, however, there has been a decrease in spectator attendance and the amount wagered, and the sport may be in decline in the USA.

Sled racing is based on the use of dogs to pull sleds in North America and Europe. It is increasing in popularity in Europe, North and South America, Australia and New Zealand, and is now carried out even in countries with little snow, where the sleds are wheeled. There are different types of racing and the sled and the teams vary in size and races vary in length. Many different breeds of dog and crossbreds are used in sled races but traditionally husky types are preferred for marathon races across snow and ice. The marathon races on snow in North America are of particular concern with regard to the welfare of the dogs. In this chapter, the welfare of greyhounds and sled dogs will be discussed.

2.   GREYHOUND RACING

Greyhound racing developed from the sport of coursing greyhounds after hares initially in open-field coursing but more recently in closed-field coursing. In open-field coursing, which is now banned in some countries, two greyhounds were set to chase a hare across open countryside. This sport can result in serious injuries to the dogs as the ground may be uneven and they may be injured going through fences. To protect the dogs and make the competition more even, closed-field coursing was developed. In this type of coursing, wild hares are captured and trained to run the length of a field and to escape through a brush fence. During a coursing match a hare is released into the field and two greyhounds are walked up towards it. When the hare is sighted the dogs are released and they chase the hare down the field. The hare usually escapes and the greyhounds are judged on the way they run the hare. Closed-field coursing is banned in some countries because of concerns about the welfare of the hare not the dog. In closed-field coursing, the owners do not want the hare to be caught as this adds to the likelihood of the dogs being injured and as the winning dogs continue to race in a series of heats during the day a kill may tire the dog and reduce its chances of winning in future.

The development of oval race tracks and an artificial hare in the early 20th Century made greyhound racing a popular sport both for participants and spectators. It is particularly popular in England, Ireland, the USA, Australia and New Zealand, but also in Macau, Mexico and Indonesia (Guccione, 1998). Tens of thousands of greyhounds are registered to race in the USA, Ireland and Australia each year (Table 1) and there is a major trade of greyhounds from Ireland to England and from Australia to Asian countries.

Greyhound racing is controlled by national organisations in Ireland and the UK. In the UK, however, many tracks, sometimes called flapper tracks, are not controlled by the national organisation. Greyhound numbers are decreasing in the USA (Table 2), probably as a result of the development of other methods of gambling.

Table 1. Number of greyhound puppies born and dogs registered to race in the countries where greyhound racing was most popular in 2001* or 2002**.

Country	Puppies born	Number of litters	Pups/litter	Dogs registered to race	Fate unknown#
USA*	Not available	5,015	6	26,797	5,901
Ireland*	Not available	3,731	8	20,694	9,000
Australia**	18,240	3,040	6	12,875	5,365
Britain*	5,500	696	8	4,165	1,400
New Zealand**	648	108	6	531	117

The number of puppies not accounted for by racing
Data adapted from the greyhound association web-sites for each country

Table 2. The number of litters born, puppies born, dogs registered to race, adopted dogs, dogs retained for breeding, dogs killed after racing and puppies killed, from 1991 to 2001 in the USA.

Year	Litters born	Puppies born	Registered to race	Adopted	Breeding	Racing Dogs killed	Puppies killed
2001	5,015	32,698	26,797	13,000	1,800	11,997	5,901
2000	5,234	34,126	26,464	13,000	2,000	11,464	7,662
1999	5,266	34,334	27,059	13,000	2,000	12,059	7,275
1998	5,034	32,822	26,036	13,000	2,000	11,036	7,686
1997	5,192	33,852	28,025	12,500	2,000	13,535	5,827
1996	5,438	35,456	28,877	12,000	2,000	13,977	6,579
1995	5,749	37,483	31,688	10,000	2,100	19,588	5,795
1994	6,232	40,633	34,746	8,500	2,200	24,046	5,887
1993	6,805	44,369	39,139	6,000	2,500	30,639	5,230
1992	7,690	50,139	38,023	3,000	2,592	32,523	12,116
1991	8,049	52,479	38,430	1,000	3,000	33,930	14,049

Data adapted from the greyhound association web-site for the USA

Racing greyhounds and show greyhounds are from different bloodlines. Racing greyhounds have to be physically sound. In general as only successful racing dogs are bred the breed suffers from few of the hereditary skeletal problems which beset other breeds. They have a placid temperament and seem to hold their excitement for training and racing. They are rarely aggressive to humans and other dogs but may chase and kill cats and livestock. Greyhounds that are aggressive to other greyhounds during racing are banned from racing and are not bred from.

Veterinarians attend greyhound racing and may stop a dog from racing, or prevent racing if conditions are bad for the greyhounds' welfare. However, there are problems with the independence of veterinarians. In the USA, track veterinarians are employed by the state and are independent (Poulter, 1996) but, in the UK, veterinarians may be employed by the track and come under pressure if their opinions regarding the fitness of a runner or the safety of the track differ from that of track management (Sweeney, 1996).

3.   TRAINING GREYHOUNDS AND WASTAGE

The training of young greyhounds varies and three different methods are described in Bloomberg *et al.* (1998). In the USA, Koerner (1998) recommended a lot of handling and leash training when the puppy is 3–4 months of age, lure and crate training at 8 months of age, and starting athletic work at 13–14 months, gradually moving onto the track. In Ireland, Cuddy and Dalton (1998) recommended that breaking-in should start between 14 and 16 months of age, when dogs are lure and crate trained, and athletic training begins. In Australia, Stephens (1998) recommended full athletic work at 16 months of age. Training using live animals as lures is forbidden and may be counterproductive, but Koerner (1998) used coyote skin lures. The athletic training protocol will vary from dog to dog but generally involves an increasing workload. Walking strengthens the bones and allows remodelling to facilitate the changes that occur during sprint training (Staaden, 1998). If dogs are trained too quickly they may be injured and this may be a major cause of loss between weaning and racing.

The loss of puppies between birth and weaning is not documented and the loss from weaning or ear tattooing to registration for racing is also difficult to determine (Tables 1 and 2). However, as the number of puppies produced in the USA decreases, the number of puppies that disappear between birth and racing registration also decreases both in real numbers and also as a percentage of puppies born. This suggests that the rearing and management of young dogs has improved. The number of dogs killed for physical or behavioural reasons during initial race training is not available. It is likely that a small percentage will not follow the lure, not be fast enough to qualify for racing, or get injured and be killed. Greyhounds usually race until they are about 2.5–3.5 years of age. A few dogs will race until they are 5 or 6 years of age but these are exceptional. On retirement from racing the majority of dogs are killed, despite having the potential to live to 10–14 years of age.

4.   INJURIES SUSTAINED DURING RACING

Greyhounds sustain particular musculoskeletal system injuries during training and racing which are not common in pet dogs (Hickman, 1975; Prole, 1976). These include fractures of the accessory carpal, metacarpal, metatarsal, central tarsal, fourth tarsal, and proximal, middle and distal phalangeal bones; and instability of the proximal, middle and distal phalangeal joints (Anderson et al., 1995a). In two surveys of greyhound racing tracks in Florida, USA, injuries identified at the tracks were monitored between 1984 and 1990 and again in 1990–1995 (Table 3). The majority of injuries were to the tarsal (hock) joint but the percentage of this injury decreased from 52.2% in the first survey to 44.3% in the second, while the number of foot injuries (metatarsus, metacarpus) increased from 8.3% to 15.4% (Bloomberg & Dugger, 1998). These authors speculated that this change might be due to increased banking at the turns and faster racing surfaces making the stress forces more vertical on the bones of the foot, in contrast to the torsion placed on the bones on flatter tracks. Thus, changing the design of the track and running surfaces may change the type of injuries sustained. The change from grass to sand track has reduced injuries to toes (Poulter, 1981), although they still remain a common injury.

Table 3. Number (and percentage) of greyhound injuries at six and 16 racing tracks in Florida, between 1984–1990 and 1990–1995, respectively.

Six race tracks, 1984-1990		Sixteen race tracks, 1990-1995	
Injury	**Total (%)**	**Injury**	**Total (%)**
Tarsus	397 (52.2%)	Tarsus	562 (44.3%)
Muscle	85 (11.2%)	Muscle	125 (9.85%)
Foot	63 (8.3%)	Metatarsus	167 (13.2%)
Toe	63 (8.3%)	Toe	99 (7.8%)
Foreleg	46 (6.0%)	Metacarpus	28 (2.2%)
Other	58 (7.6%)	Laceration	31 (2.5%)
		Cramping	10 (0.8%)
		Long-bone fracture	78 (6.2%)
		Carpus	63 (4.96)
		Other	106 (8.4%)

Adapted from Bloomberg & Dugger (1998)

Injuries during racing are due to many factors, including speed, fitness of the dog, length of the race, character of the track, and the weather. A comparison of five racing tracks in the USA found that the injury rate over the years varied from 3.4 to 5.4% that is about 4% of all races by greyhounds resulted in injury (Sicard *et al.*, 1999). There was no obvious seasonality to the incidence of injuries and no correlation between injury rate and ambient temperature or the dog's bodyweight. The design of the track influenced the rate of injury and one track had a significantly higher incidence of injury than the other four. On that track speeds were the highest and the turns were banked more to increase speed. Higher-grade races resulted in more injuries than lower-grade, maiden and schooling races (Table 4), and in some race lengths there were more injuries. However, the race number and dog's starting position (Bloomberg & Dugger, 1998) had no significant effect on injury rate, although dogs in Starting Position 4 were most often injured and there were many injuries at the first bend as dogs bumped each other (Sicard *et al.*, 1999). When the location at which the injury occurred was known it was most often sustained at the first bend when speed was high and there was maximum congestion. This is similar to the results of the study in Florida, USA, where nearly half the injuries (471/830) in 5/16 mile races occurred at the first bend and nearly a fifth (153/830) at the second bend (Bloomberg & Dugger, 1998). Race length has a significant effect on the rate of injuries in Florida, and 3/8 mile races had nearly twice as many injuries as 5/16 mile races, due perhaps to the dogs having to negotiate an extra turn (Bloomberg & Dugger, 1998).

Table 4. Injury rate in races of different grade and length, from five racing tracks in Wisconsin, USA.

Grade	Injury rate	Distance	Injury rate
Grade A	5.5%	3/16-mile	7.7%
Grade B	4.9%	5/16 –mile	4.3%
Grade C	4.2%	3/8-mile	4.7%
Grade D	3.9%	7/16-mile	7.2%
Grade E	3.7%		
Maiden races	2.7%		
Schooling	0.9%		

Adapted from Sicard *et al.* (1999)

Greyhounds are subjected to centrifugal, gravitational and frictional forces when racing (Ireland, 1998). To counter these, greyhounds lean in to the left side as they corner whilst running anti-clockwise. The dog nearest the left side has to lean more than the dogs further out to the right, in order to maintain its position. To reduce the centrifugal forces acting on it the greyhound can either slow down or race on a larger radius (Ireland, 1998). If it slows, the dogs following it will be interfered with and if it goes for a larger radius it will run out in front of the other dogs. To reduce centrifugal force the track can also be changed by increasing the radius of the turns and the steepness of the banking at the turns. However, making these changes to the track may increase the speed and then as the speed of racing increases the injury rate may again increase (Sicard *et al.*, 1999).

More than 90% of fractures of the central tarsal bone occur in the right leg (Boudrieau *et al.*, 1984a) because of the pressure placed on this joint by counter-clockwise racing (Anderson *et al.*, 1995b) when the burden of micro-damage to the bone overwhelms its micro-elastic properties (Lipscomb *et al.*, 2001). Many greyhounds are killed or retired after fracturing this bone. However, the prognosis for a return to racing is good after surgical fixation, and 88% of 81 dogs in one survey returned to successful racing (Boudrieau *et al.*, 1984b). In one study of eight greyhounds with fractures of this bone, all returned to racing within 12 months and owners reported excellent results from six dogs, a good result from one, and one dog was retired (Guilliard, 2000). Dogs with a fracture of this bone are often not treated because of the cost involved, the long rest period, and the assumption that the dog will not regain pace (Guilliard, 2000).

Injuries to greyhounds are common and successful treatment generally means a return to racing. Many owners would wish for a speed equal to those reached before injury but dogs can be reclassified in some racing programmes and compete in lower-grade races. Many injuries can be treated successfully if a diagnosis is made quickly, treatment is rapid, and the dogs are allowed time to rest to recuperate. The duty of the veterinarian at official greyhound races is to ensure the welfare of the dogs. Dogs injured on the track are treated if possible or the veterinarian will advise and initiate treatment if necessary.

The common injuries of coursing greyhounds are different from those of racing greyhounds. Injuries involving the carpus and tarsus of racing greyhounds do not occur so frequently in coursing dogs, which appear have more injuries involving the distal limb (Brown, 1998).

5.   ADOPTION OF GREYHOUNDS

The greyhound is an ideal pet, and can live for 10–14 years. It is healthy, short-haired and has few hereditary diseases. Those that have raced may have some chronic injuries but these are generally not significant if the dog is no longer racing. Greyhounds are categorised behaviourally as having low aggression, low reactivity and low immaturity (Bradshaw *et al.*, 1996) and so should make friendly and inactive pets. In the past, greyhounds not required for breeding were killed when their racing career was finished, due to the perception that they were aggressive animals. They do have a tendency to chase anything that runs from them so cats may be endangered should they run away. Greyhounds can be trained not to chase cats or small dogs.

In the last decade, more and more dogs were adopted after racing (Table 2) (Dee, 1998) and the greyhound industry worldwide has supported this move. In the USA, about 60% of racing greyhounds were adopted (Anonymous, 1995). These adopted dogs may retain some of the injuries sustained during racing, such as injured toes, some degenerative joint disease, and old injuries to the carpal and tarsal joints plus some muscle damage. Bitches may have an enlarged clitoris due to having been given testosterone to suppress oestrus. Adopted greyhounds lose muscle mass over time and probably recover from the stress-related syndromes seen in some dogs during racing. The euthanasia of many greyhounds when they have finished racing is unpalatable to many, but if the animals are killed painlessly and without stress, then it is not a welfare issue.

6.   RACING STRESS IN GREYHOUNDS

Successful racing dogs and competing dogs work in a highly competitive environment and when an animal fails to cope with this environment it may become stressed. The success of racing dogs can be reduced by injury, ageing or being worked in different or difficult circumstances. If no medical or surgical reason for poor performance can be identified then it may be due to stress. Little has been published in the scientific literature on stress in racing dogs but heavy exercise is an extreme stressor. During a 400 m race, the temperature of a racing greyhound may rise to 42°C, its PCV to 70%, its blood pH decline to less than 7.0, and cardiac output increase up to several times the normal level. This gross change in the physiological status of the greyhound would become fatal if continued (Holloway, 1998).

Immediately prior to a 704 m race, there were major physiological changes, *viz* plasma volume decreased by 10% and total circulating RBCs increased by 60% (attributed to increase in the number of cells *per se*), the blood volume increased by 24% and PCV by 29%. Five minutes after the race, the plasma volume was 20% below resting value and total RBC had increased by 73% resulting in a 40% increase in PCV (Toll *et al.*, 1995). The physiological changes in the greyhound before and during exercise enhanced blood volume and aided acid/base homeostasis, both necessary for sprinting (Toll *et al.*, 1995). The greyhound does not have the ability to release large numbers of RBCs from the spleen (Rose & Bloomberg, 1989) but has an extremely high resting haematocrit, possibly the highest of any mammal.

The length of a race influences the physiological changes which occur within the greyhound. Plasma lactate ion concentration, potassium ion concentration, haematocrit, and total protein increased with race length from 402 m to 503 m to 704 m (Nold *et al.*, 1991). The increase in plasma lactate ion concentration and the decrease in arterial pH were significantly greater than in horses or humans exercising similarly. This severe metabolic acidosis caused by increased plasma lactate concentrations peaked five minutes after a 400 m race but had returned to normal after 30 minutes (Rose & Bloomberg, 1989), demonstrating a great capacity for lactate production and metabolism in the greyhound. Glycogen breakdown in muscle appears to be higher in the greyhound than in racing Thoroughbred horses (Dobson *et al.*, 1988).

Three hours after racing 722 m, the vital signs (heart rate, respiratory rate, temperature) and blood-gas and acid-base values in greyhounds had returned to normal (Ilkiew *et al.*, 1989). However, many haematological values were still different from baseline values, with high values for WBCs and neutrophils, and low values for PCV, total plasma protein, haemoglobin, RBCs and lymphocytes. Plasma creatine kinase (CK) concentrations were still high also (Ilkiew *et al.*, 1989). The increase in numbers of WBC is a normal stress response associated with release of cortisol (Rose & Bloomberg, 1989).

Because the greyhound is so severely tested physically during racing, minor stressors, either emotional or physical (Table 5), can affect the ability of the dog to maintain physiological homeostasis, and if the animal continues to be raced it may result in the development of physiological exhaustion (Holloway, 1998). During this process there will be a loss of form, suppression of immune function, and specific stress-related syndromes may develop. Any one of several physical or environmental factors such as subclinical disease and parasitism, injury, poor nutrition, lack of water, poor kennel environment and poor handling may be sufficient to cause stress. Emotional factors may also be important and the competitive drive of most racing dogs or excessive pre-racing excitement might be sufficient to cause stress. Excessive training, especially in hot humid conditions, may cause distress. There is little information on the effect of different training schedules on the physiological responses of greyhounds to racing (Rose & Bloomberg, 1989).

Racing greyhounds need to be fed a highly digestible, high carbohydrate, low fat, moderate protein diet. Food has to be given more than four hours before training or racing and they need a high carbohydrate snack 30 minutes after racing, to enhance glycogen repletion (Todd & Reynolds, 2000). Inadequate feeding may result in poor performance (Toll, 2000a) and stress.

Table 5. Causes of stress in the competitive dog, including racing dogs.

Environmental	Emotional
Overcrowding	Fighting
Poor sanitation	Pre-race excitement
Infections and infestations	Transport
Poor nutrition	Competition for food
Poor husbandry	Need for exercise
Racing injuries	Sexual competition
Temperature and humidity	Overwork
Poorly maintained kennels and yards	
Iatrogenic injury	
Poor vaccination and quarantine practice	

Adapted from Holloway (1998)

Poor performance in greyhounds may be related to severe stress and affected dogs can develop dry seborrhoea and dehydration, rhabdomyolysis, non-regenerative anaemia and a loss of body fat and muscle mass (Holloway, 1998). Holloway (1998) proposed that two syndromes seen in racing greyhounds, polyuria/polydipsia (also called water diabetes syndrome) and balding thigh syndrome were stress-related. Balding thigh syndrome, however, may result from rubbing on crates.

7.   SLED RACING

Sled racing has become popular in the last decade. In North America, there are at least three types of sled racing, including (1) sprint racing, which consists of three races of 5–20 miles carried out over three days; (2) long-distance races, in which dogs race for days or weeks; and (3) stage races, where each stage is 40–80 miles per day and there are 10–20 stages in all. A variety of dogs are used but the usual ones are Huskies (Grandjean *et al.*, 1998). In Scandinavia, one to four dogs pull a small sled called a pulka, followed by a person on skies. These pulkas are often pulled by German Shorthaired Pointers which are very fast. In 1998, there were sled teams in 31 countries (Table 6) which shows the popularity of this sport even in countries with little snow. In those countries, the sleds often have wheels.

Table 6. Countries which are members of the International Federation for Sled Dog Sport.

Country	Members	Events	Races	Teams
Andorra	5	1	2	4
Argentina	27	1	1	18
Australia	200	15	22	152
Austria	600	12	66	1,151
Belgium	250	6	24	480
Canada	6,200	122	321	4,591
Chile	13	1	1	8
Czech Republic	80	10	28	345
Denmark	130	16	75	460
England	550	16	52	1,084
Finland	1,000	14	37	464
France	1,600	15	69	1,744
Germany	2,400	48	284	3,161
Holland	200	6	30	481
Hungary	30	2	2	20
Ireland	3	0	0	3
Italy	400	30	137	2,390
Japan	345	7	47	1,329
Luxembourg	0	1	6	44
New Zealand	127	11	49	336
Norway	800	90	193	1,966
Russia	80	4	8	70
Scotland	120	6	24	381
Slovenia	30	1	3	25
South Africa	100	10	10	82
Spain	250	4	26	220
Sweden	5,000	89	289	3,977
Switzerland	175	9	49	671
Ukraine	5	1	1	5
United States	11,5000	330	1,097	10,525

Adapted from Grandjean (1998)

Sled dogs trained for racing have increased PCV values and RBC counts at rest and in response to exercise have lower heart rates and rectal temperatures than unfit dogs (Burr *et al.*, 1997). Dogs are trained for long-distance races over a period of six months and their diet is adjusted throughout this period. During training dogs have major cardiovascular (Constable *et al.*, 2000) and other physiological changes to allow the extended exercise. During long distance races dog may cover 100 miles each day for 8–12 days (Lee *et al.*, 2004) which is associated with a very high metabolic rate and the use of more than 12,000 kcal per day. This is extremely strenuous activity, often in very cold conditions, and dogs may be affected adversely by the effort.

8.   THE STRESS OF SLED RACING

The stress and distress caused by sled racing will vary depending on the type of race the dogs are involved in and the climate and terrain over which it is run. There are specific injuries associated with racing sled dogs. These include injuries to the pads, web or nail beds of the foot, and other types of lameness due to a variety of injuries, especially the carpal joint and tendons of the foreleg (Grandjean *et al.*, 1998). Frost-bite may occur when exposed skin and areas made hairless by chafing of booties and harness cool to 0°C. Hypothermia occurs due to low ambient temperatures and wet conditions. Resting dogs are more prone to hypothermia than running animals and some breeds such as the poodles have been banned from marathon races because of their poorly indulated coats. Sled dogs sleep in a ball to reduce heat loss. Rectal temperatures below 37.8°C are indicative of hypothermia, and if the core temperature dops below 35.5°C the dog is in critical condition and needs urgent treatment (Grandjean *et al.*, 1998). Hyperthermia occurs during warm-weather racing.

Short-distance races using wheeled sleds in warm climatic conditions will cause different problems than long-distance races run over ice and snow, such as the Iditarod race in Alaska. This race is 1,771 km long, over snow and ice tundra from Anchorage to Nome. It takes place every March and temperatures range from –34 to +1°C (Davis *et al.*, 2003b). Teams usually take between 9 and 14 days to complete the race. Most participating dogs weigh between 18 and 29 kg and are 2 to 10 years of age (Davis *et al.*, 2003b). Most participanting dogs are 5 to 7 years of age. It is regarded as one of the toughest endurance races for animals and humans. Each year about 60–70 mushers (drivers), with 16–20 dogs each, start the race. The race is overseen by veterinarians. The dogs are examined at 27 checkpoints and those that are injured, exhausted or require treatment are flown back to

Anchorage (de Jong, 1992). The dogs are fed low volume and high nutritional value diets and need two to four times maintenance to maintain bodyweight and to race. During this race, injuries to the feet are common and frost-bite to the prepuce, scrotum or mammary glands of recently weaned bitches, occur.

In long-distance sled races such as the Iditarod, up to a third of the dogs do not complete the race and the reasons cited are usually injury, fatigue, lameness and diarrhoea, and sometimes veterinarians are concerned about dehydration. Dehydration was not seen in finishing and non-finishing dogs in the 1991 Iditarod race (Burr *et al.*, 1997) and probably is not a common disabling condition in sled dogs. The high fat diet fed to sled dogs provides metabolic water, and dehydration may occur when the tired dog refuses to eat because of stress or diarrhoea. If dehydration is greater than 7% the dog is not allowed to continue racing (Grandjean *et al.*, 1998). In a study of the 1993 Yukon Quest race, there were no indications of dehydration, significant plasma electrolyte abnormalities or myopathy in dogs retired from the endurance race (Hinchcliff, 1996). The sex and age of the dogs that completed the race or were retired were similar and of the 15 dogs retired, 13 were lame, two had diarrhoea, one had fever and only one was dehydrated. Dogs are retired by the driver after discussion with a race veterinarian and officials. There is need for further investigation into the causes of dogs being retired from races.

Dogs finishing the 1995 Iditarod race had lost a maximum of 9% bodyweight over the course, but the CK levels of these dogs were not clinically significant (Hinchcliff *et al.*, 1998). In that study, the dogs that had been retired before they had completed 500 miles had more clinically significant CK levels, indicative of exercise-induced rhabdomyolysis, than dogs retired after 638 miles or which finished. This condition is apparently common in long-distance sled races (Grandjean, 1998) and is known to be fatal to dogs during the first 400–500 miles of long races. Individual dogs may be particularly susceptible to this condition during the first part of a long race. In the 1995 race, there was no relationship between the speed with which the race was run (finishing order) and bodyweight loss or biochemical values of dogs (Hinchcliff *et al.*, 1998). The weight loss was probably insignificant as dogs were weighed about 4 hours before the race began and had just received food and water. This indicates that dogs receive adequate food and water to meet the demanding requirements of this race. There was also no dehydration observed, which means that the dogs received more than the five litres of fluid per day that they required.

Non-finishers of the 1991 Iditarod race had higher post-race CK and aspartate transferase (AST) activities than finishing dogs, and though the difference was not significant overall this was probably due to the great variation between dogs. Burr *et al.* (1997) suggested that the higher levels were probably associated with muscle damage. This opinion was supported by a post-mortem survey of dogs that died suddenly during long-distance sled races, which found severe muscle necrosis similar to the lesions caused by rhabdomyolysis in some dogs (Burr *et al.*, 1997). They suggested that CK and possibly AST could be used to monitor fatigue in long-distance sled races. The number of non-finishers in this study was eight of a team of 17, and three of these were lame while five were assessed as fatigued. CK activity has high specificity (83%) and low sensitivity (32%) for muscle damage. This means that dogs with high plasma CK activities are likely to have muscle damage, but those with low levels may also have damage but are false negatives (Aktas *et al.*, 1993).

The antioxidant mechanisms of under-trained sled dogs may be inadequate to cope with the antioxidant requirement of endurance racing (Hinchcliff *et al.*, 2000). Such racing uses a great deal of energy and results in considerable release of free radicals and oxidant stress. This may result in lipid peroxidation (Piercy *et al.*, 2000) and muscle damage during racing, which can be assessed by plasma CK activity. However, feeding supplements of antioxidants, such as vitamins E and C and beta carotene, to exercising sled dogs did not reduce exercise-induced increases in plasma CK (Piercy *et al.*, 2000) but may have reduced oxidative damage (Baskin *et al.*, 2000). The damage to muscles sustained by racing sled dogs may be due to mechanisms other than oxidative stress (Piercy *et al.*, 2000).

Gastric ulcers are reported to be a problem in racing sled dogs and may be a cause of sudden death. At the end of the 2001 Iditarod race nearly half (34/70 or 48.5%) the dogs examined had gastric lesions, which may have been an underestimate as many dogs had food, foam or bedding in their stomachs which could have prevented observation of lesions (Davis *et al.*, 2003b). This is a very high percentage of dogs with gastric lesions and Davis *et al.* (2003b) speculated about their causes. Gastric ulcers may be due to a high fat diet (70% fat), which delays stomach emptying and leads to hyperacidity. Other possible causes include bacterial infection (*Helicobacter* spp), trauma caused by cold foreign bodies or snow, cyclo-oxygenase inhibitors, or exercise-induced visceral ischaemia and stress. Many dogs without lesions had foreign bodies in their stomachs and drugs were unlikely to be a cause. Exercise-induced ischaemia was unlikely to be a cause as sled dogs maintain visceral perfusion during exercise.

Gastric ulceration is also common in human athletes undertaking vigorous exercise and Davis *et al.* (2003b) concluded that the stress of endurance racing with elevated plasma cortisol concentrations and exogenous glucocorticoids may predispose sled dogs to gastric ulceration. When omeprazole, a proton pump inhibitor, was given to sled dogs before and during the Iditarod race the severity of gastritis in the treated dogs was reduced compared with control dogs (Davis *et al.*, 2003a). The prophylactic use of omeprazole may improve the welfare of sled dogs by reducing the severity of subclinical gastric disease but further work is required to understand fully the aetiology of this condition.

Sled dogs in races such as the Iditarod have very high metabolic rates, with energy expenditures of 10,000 kcal per day (Hinchcliff *et al.*, 1997a). Tests showed that plasma thyroxine (T_4), free T_4 and tri-iodothionate (T_3) concentrations were decreased after prolonged sub-maximal exercise in trained sled dogs and some dogs had levels below that seen in normal pet dogs (Panciera *et al.*, 2003; Evason *et al.*, 2004). Thus, training appears to effect thyroid gland function on middle and long distance trained sled dogs. However, lack of training in the three months after the Iditarod race did not change thyroid gland test results from those levels seen immediately after the race (Lee *et al.*, 2004). Levels of T_4, free T_4 and thyroid stimulating hormone decreased significantly during the Iditarod race itself but only T_4 was lower in dogs that completed the race than those withdrawn during it (Lee *et al.*, 2004). It may be that thyroid hormone levels in sled dogs of the husky type are lower than other breeds (Lee *et al.*, 2004) but prolonged exertion in subfreezing conditions increases metabolic rates in sled dogs (Hinch-cliff *et al.*, 1997a). Sled dogs are usually fed a fat-rich diet to maintain energy balance but an inadequate energy intake could alter thyroid hormone metabolism. The cause and significance of changes in thyroid hormone levels in racing sled dogs need to be investigated further.

Sled dogs need to be fed at least four hours before training or racing, to empty the bowels and prevent stress-induced diarrhoea (Reynolds, 2000). They need a highly digestible high fat, low carbohydrate, moderate protein diet (Toll & Reynolds, 2000) and are usually fed commercial food plus a fat source. During intensive training, a diet with 35% dietary protein as energy is required (Reynolds *et al.*, 1999). Dogs in training fed 18% dietary protein as energy sustained significantly more soft tissue injuries than dogs fed more protein. PCVs were also higher after dogs were fed the higher protein diet for 12 weeks.

Prolonged running is associated with decreased serum sodium and potassium concentrations in sled dogs (Hinchcliff *et al.*, 1997b). This is not due to thermoregulation by sweating but may be due to a high renal solute load mandated by the large energy intake and expenditure by these dogs (Hinchcliff *et al.*, 1998b). The clinical significance of these decreases is unknown but a significantly greater proportion of 151 dogs that retired from the 1995 Iditarod race had low sodium concentrations compared with dogs that finished, so it might be involved in the fatigue syndrome.

Strenuous short-term sled pulling resulted in priming and activation of platelets and of neutrophils in Siberian Husky dogs (Moritz *et al.*, 2003). Activation of platelets may be preparation for the wounds which can occur during predation but which may make dogs succeptible to thrombotic disorders (Moritz *et al.*, 2003). Neutrophil activation has also been reported in human marathon runners; and while it may be useful in infectious conditions, adverse effects are also possible. Activated neutrophils are rigid and may lodge in small vessels and cause death of parenchymal cells and organ failure. The significance of platelet and neutrophil activation in sled dogs remains unclear.

The welfare of sled dogs during training and racing may be compromised in many ways. The number of dogs that suffer fatigue during the racing suggests that the animals are being pushed to the limit of their ability. In one study, a sample of 261 dogs was followed during the 1995 Iditarod race and 151 of these were retired from the race after a median run of 474 miles (Hinchcliff *et al.*, 1998). Many of those dogs would have been retired for fatigue, which suggests that something in the selection, training, preparation, nutrition and/or driving of these dogs was inadequate. Sled dogs used for endurance races such as the Iditarod race cannot be driven. Individual animals may be injured and a dog died during the race in 1999, 2000 and 2001. There is a very high media interest in the Iditarod race and good veterinary coverage. The race is carefully managed by experienced personnel to ensure that the welfare of the dogs is as good as the essence of the race allows. If roughly half the dogs are retired from this race which has excellent management, there have to be concerns about the welfare of dogs in races that are not so well managed, both in North America and elsewhere.

9.   MANAGEMENT AND TRANSPORT OF RACING DOGS

Some aspects of management make racing dogs especially prone to particular problems. Internal parasites are common in racing dogs held in large kennels because their exercise areas are often heavily contaminated with parasite eggs and larvae. Parasite control is an important aspect of the management of these dogs. Racing dogs spend a lot of time in the company of dogs from other kennels and this may expose them to kennel cough, so they have to be vaccinated against this disease.

As males and females race together it is important to suppress signs of oestrus, otherwise male dogs lose interest in racing and may fight, and the affected female loses form and may not recover it. Racing greyhounds have a short racing career and owners cannot afford to have bitches out of the running for the duration of two oestrus periods a year. Therefore, suppression of oestrus by pharmacological agents is common. The drugs used may have side effects including an enlarged clitoris and changes in temperament. However, as all dogs must be drug free on race day these drugs cannot be given within defined periods of racing. In the UK, testosterone and first and second generation progestogen preparations may be used (Prole, 1996).

Racing dogs are usually transported in trailers and vans especially designed or modified for their transport. Ventilation is important especially if there are several dogs in one compartment and the weather is hot and humid. The stress response of greyhounds to short-duration air transport (2 x 35 minutes) was not different when they were held in narrow (1,220 x 813 x 384 mm) or wide (1,220 x 840 x 800 mm) pens (Leadon & Mullins, 1991), but there may be a difference depending on which hold within the aeroplane the dogs were kept. There was great individual difference in the responses of dogs to transport. The effect of long-distance road travel on sled dogs going to and from races, and greyhounds being exported overland is unknown.

10.   DOPING DOGS

Greyhound and sled racing is very competitive and gambling is an important factor especially of the former. Drugs can be used to enhance, reduce or normalise behaviour (Table 7). Pharmacological enhancement of racing performance may make the dog more prone to injury. Pain relief may also make dogs prone to injury during training or racing. Doping controls are enforced at all regulated greyhound race meets and since the mid 1990s increasingly in sled races, especially the major events (Craig, 1998).

Table 7. The use of drugs in performance modification.

Modification	Drug type
Enhancement	Ergogenic aids called stimulants or 'goers'
Reduction	Depressants or suppressants called 'stoppers'
Normalisation	To stop normal physiological responses to excitement or minimise pain and discomfort

Adapted from Gannon (1998)

11.   CONCLUSIONS

Greyhound racing is an intense physical activity with attendant risks of serious injuries. The dogs are managed well and it is necessary that they are trained and fed correctly and healthily if they are to compete successfully.

Their welfare is compromised principally by the injuries sustained during training and racing. Sled dogs may have their welfare compromised during endurance races when they are physically extended over long periods of time and great distances. More research is needed to reduce injuries to all racing animals and to increase the percentage of dogs completing sled races. However, wastage is always going to be a problem in these dogs as they are competing at a high level of physical demand.

Chapter 9

THE DOG AS A RESEARCH ANIMAL

Abstract: The morphology and physiology of the dog makes it a useful model for medical and dental research. The recent clarification of breed genotypes and the existence of breed-specific diseases make the dog a particularly suitable model for research into inherited diseases common to humans and dogs. It is likely that more dogs will be used in biomedical research in the near future. It is important that the welfare of dogs used in research is optimal for the dog's sake and also because animal models of disease work best if the subjects are healthy and not stressed. Many dogs held in laboratories are probably chronically stressed. Improved breeding and training programmes, and better housing and management are required if laboratory-based dogs used in research are to have good standards of welfare. This will require improvements in social enrichment and physical facilities.

1.   INTRODUCTION

The dog is similar to the human in some ways and this makes it an appropriate animal for biomedical research. The dog is also a companion animal and there is much pressure to reduce its use as a laboratory animal and encourage researchers to improve their management (Hubrecht, 2004). The dog may be used as a research animal inside or outside the laboratory. Research conducted on dogs that live in their own homes is typically clinical, nutritional or behavioural.

The dog was an important animal in early physiological and anatomical research. In the 17th Century, Harvey used dogs to study heart movement and Marcello Malpighi used them to investigate lung function (Gay, 1984). Dogs were used by Sir Christopher Wren to demonstrate the intravenous administration of drugs. Pavlov identified classical conditioning by observing the responses of laboratory dogs. The dog played a part in Banting and Best's research into the role of insulin in diabetes mellitus (Gay, 1984). The use of dogs in biomedical research has always been a cause of controversy, and in 1903 Dr William Bayliss of the University of London won a court case against the anti-vivisectionist Stephen Coleridge who had accused the former of cruelty to a brown dog. Research on the brown dog had led to the discovery of secretin, a significant breakthrough in digestive physiology. A statue of this dog was erected by those opposed to the use of dogs in biomedical research (Jones, 2003).

Dogs are used in many branches of biomedical research. These include physiology, toxicology, surgery, dental health, and hereditary diseases of humans. They are also used in research on the health, nutrition and behaviour of the dog itself. The genetic differences, between different breeds of dog (Parker *et al.*, 2004), make them ideal populations in which to study the origins of genetic disease in dogs, and in turn simplify the investigation of corresponding genetic diseases in humans (Pennisi, 2004). Thus, the use of the dog in biomedical research is likely to increase in the near future. In this chapter, the use of dogs in research and the significance of living conditions on the welfare of dogs used in laboratories will be discussed.

2.   THE USE OF DOGS IN RESEARCH

Dogs are more expensive to keep than rodents and therefore their use in scientific research is limited to those areas where they are more appropriate than rodents. The use of dogs as research animals are detailed in scientific publications from many fields of biological research, but dogs are used a lot in research into pharmacology, dental health, toxicology or surgery (Table 1). Dogs other than beagles are used in research, especially in surgical and dental research. In an analysis of one scientific database, there was an increase in the number of scientific publications, describing research using Beagles. In 1983 there were 83 publications and this has grown to 237 in 1993, but the number has remained static since then (Table 1). There has been a decrease in the number of scientific procedures performed on dogs in the United Kingdom. In 1989, 12,625 were conducted and this dropped to 9,085 by 1992 (Hubrecht, 1995a).

Dogs are used as models for human diseases such as diabetes mellitus, ulcerative colitis and cardiovascular disease, and for surgical research such as open-heart surgery and organ transplantation. Dogs may also be suitable models for other diseases, such as benign prostatic hypertrophy, glomerulonephritis, auto-immune haemolytic anaemia, lymphocytic thyroiditis, Von Willebrand's disease, chronic pancreatitis, haemorrhagic pancreatitis, endocarditis, diabetes mellitus, narcolepsy, hydrocephalus, tetralogy of Fallot, idiopathic epilepsy, and neoplasia (Kohn, 1995). However, the recent determination of genetic differences between different breeds of dogs (Parker *et al.*, 2004) and the recognition that individual breeds are defined populations will accelerate the understanding of canine diseases of genetic origin, and also be a key tool in understanding the genetic basis of diseases which are common to dogs and humans (Pennisi, 2004). This will likely increase the number of dogs involved in research into inherited diseases common to dogs and humans. This will increase the range of breeds required for research. The standards of management developed for beagles and the knowledge gained about their requirements will need to be reassessed to cope with other breeds.

Table 1. The number of scientific publications in which Beagles were used as research animals, or dogs were mentioned according to a scientific database.

Area of research	No. dogs used		
	1983	1993	2003
Beagles			
Pharmacology	19	69	57
Toxicology	22	28	30
Physiology	12	28	16
Dentistry/oral health	18	27	41
Veterinary medicine[a]	6	40	51
Surgery	1	29	30
Cancer	5	4	4
Miscellaneous	2	2	8
Total beagles	85	237	237
Dogs	1,519	2,400	2,396[b]

[a] Includes veterinary research into the health of laboratory beagles

[b] The majority of these references are in journals of veterinary medicine and surgery

Many dogs are used for research on the health, surgery, behaviour and nutrition of dogs *per se.* This work is undertaken predominantly at facilities belonging to the major dog-food companies and veterinary colleges. Both these types of organisations are extremely sensitive to the public demand that dogs in research be managed under conditions where their welfare is maintained at a very high standard. However, in the development of new surgical techniques, even with good anaesthesia and analgesia, there may be painful consequences. Moreover, in the development of new treatments for specific diseases, dogs may be subjected to novel, but ineffective or less effective treatment protocols than are available. Animals may also be bred with particular diseases or be managed so as to become diseased in order to test new treatment protocols, as investigations into the aetiology of disease often requires animals that are suffering from the disease and that have not been treated.

Dogs are also used for research into human psychological problems and for research into the cognitive skills of dogs themselves (Cooper *et al.*, 2003). The use of the dog as a model for research into human psychological conditions such as obsessive behaviour has been promoted by some canine behavioural specialists, as a way to increase our understanding of the behavioural problems of dogs (Overall, 1997). This type of research may be invasive, with resultant brain damage, infusions to the brain, and intensive monitoring of neurotransmitter and neuronal activity being conducted. The production of dogs with the required behavioural problems may be necessary if such research is to be successful. This may involve breeding from susceptible lines and then deliberately exposing the offspring to inadequate environmental conditions in order to produce anxiety, compulsive behaviours as required.

3.   SOURCE OF LABORATORY DOGS

Dogs used in laboratory research originate from three major sources: special breeding facilities that produce dogs for research, shelters or dog pounds, or individual owners, breeders or traders. In some European countries, dogs to be used in laboratories must come from registered breeding facilities (Hubrecht, 2002). In the USA, dogs can be purchased from a United States Department of Agriculture licensed dealer or directly from a municipal pound (Dysko *et al.*, 2002). Licensed breeding establishments are monitored by designated inspectors and can be forced to maintain high standards of management in order to remain approved.

In general, research is better if the dogs used are uniform, well reared, healthy and from a known and well-defined source. If from a registered breeding facility, the research laboratory can dictate how the dogs are reared and managed prior to purchase. This is particularly important with regard to the proper socialisation of the dogs as pups to humans, which makes adult dogs safer and easier to handle.

Dogs may also be trained in the breeding facility, using positive reinforcement, to sit still while simple procedures are performed on them. These procedures might include a clinical examination, the taking of rectal temperatures, blood and urine samples, and giving tablets and injections intravenously, intramuscularly and subcutaneously. The dogs can be trained to accept restraint and muzzling as required. This training early in life reduces the stress experienced by dogs during research and adds to their value as research animals. Dogs, reared in special breeding facilities, may experience institutional lives from the start and may not suffer from loss of freedom when moved into a laboratory.

The common dog breed produced specifically for laboratory use is the Beagle, but mongrels and Foxhounds are also used frequently (Dysko et al., 2002). Beagles are physically and temperamentally suited for research. They are middle-sized, thus suited for sampling purposes, but not so heavy to make it difficult for laboratory staff to lift them onto tables and move them around. They are short-haired and thus do not lose hair that blocks drains, nor do they need grooming or clipping before blood sampling. They are quiet, gentle animals suited to living in small packs, and are not particularly aggressive. Their existence for some time as 'the' laboratory dog may make it easier for handlers, technicians and research scientists to use them without becoming too emotionally attached to them.

The health, nutrition, growth and behaviour of dogs especially produced for laboratory work can be controlled from birth until they are moved into the laboratory facility. The genetics of the animals can be determined and strains of dogs well suited to laboratory work can be bred. If healthy dogs with a good temperament are used it makes the work of the staff in a laboratory easier and dogs are likely to be better treated as a result.

When dogs are sourced from pounds and shelters their temperament is poorly defined and they may not be suited to long-term research projects. There may also be problems with their health and physiological status and they may be of an unknown age. Their previous experiences and temperament might make them difficult or even dangerous to handle, and this will affect the ease with which they can be handled. Thus, although these animals may be easily obtained and be cheap, their variability could compromise the research by either producing poor results or work that has to be repeated. Additionally, they may be more difficult for laboratory staff to work with. Many shelters and pounds will not allow their dogs to be used in research and will kill surplus or unwanted dogs rather than sell them for research.

Individual owners may surrender animals to a laboratory, and dog traders may source dogs for a laboratory, but the same problems with regard to their unknown previous history and heterogenicity might exist. However, Greyhounds retired from the race track share many of the characteristics of the 'laboratory' beagle, and with proper management may be suited to laboratory work. Dogs obtained from pounds and traders are usually mongrels or of the middle-sized breeds common to the country in which the laboratory is sited. These commonly include German Shepherd Dogs, Doberman Pinschers, and Labrador and Golden Retrievers (Dysko et al., 2002). The provision, in some countries, that all dogs must come from approved breeders can reassure owners that if their beloved dog goes missing it will not end up in a research laboratory.

The use of animals that are homogenous both genetically and with regard to their experiences before they are used in research allows for the requirement of fewer animals. Thus, the supply of dogs for research from well-defined backgrounds is important in fulfilling one of the requirements of the three Rs (reduce, refine, replace) formulated by Russell and Burch (1959).

4.   MANAGEMENT

The management of laboratory dogs must ensure that the animals are healthy, well fed, trained to accept minor procedures, and that they can behave in a generally normal manner with little abnormal behaviour. Adequate management and facilities safeguard the welfare of the animal but are also necessary for sound research. Inadequate management can impact on the results of research, making it worthless (Donnelley, 1990), due to animals being physiologically or psychologically abnormal before the research commences. It is not difficult to ensure that animals are healthy and adequately fed and trained appropriately, but whether they are behaving normally is more difficult to assess.

It is possible that dogs held in cages for extended periods of time suffer chronic stress. In an important study in the Netherlands, dogs, about six years of age, held individually for years in cages (1.7 m^2) and allowed into an outside cage (3.6 m^2) for 6 hours each day, were chronically stressed, and had urinary cortisol to creatinine ratios significantly greater than pet dogs and dogs held under less austere conditions (Beerda et al., 2000). In addition, three to four year old beagle bitches held in pairs in cages (2.4 m^2) and allowed similar exercise outdoors had urinary cortisol to creatinine ratios indicative of chronic stress. The placement of dogs in pairs rather than as individuals appeared to have little effect on the level of chronic stress. However, dogs housed individually with an indoor (2.1 m^2) and outdoor section (5.6 m^2), and walked outside on a regular basis for 90 minutes, were not as stressed as those mentioned above. These physiological results are supported by behavioural observations. The six year old dogs had high levels of paw-lifting, a behaviour found to indicate acute (Beerda et al., 1998) and chronic (Beerda et al., 1999a) stress. The behavioural responses of the older dogs and the beagle bitches to a slamming door suggested they experienced more acute stress in response to this stressor than pet dogs and those walked frequently (Beerda et al., 2000).

The behaviour of dogs caged in laboratory facilities should be as normal as possible. However, it is difficult to define normal behaviour (Chapter 11). In the simplest definition for the domestic dog, normal behaviour may be interpreted to mean the behaviour of feral dogs. This definition is not easy to use and many owned and loved dogs are not allowed to forage, live in packs, or reproduce. These behaviours are unacceptable for dogs used in laboratories. It may be best to accept that normal behaviour occurs when dogs behave in an acceptable manner and do not exhibit abnormal behaviour such as stereotypies, abnormal aggression or atypical feeding behaviours. However, comparative studies involving physiological and behavioural responses, such as those of Beerda *et al.* (2000), are more valuable in that in these studies the significance of an animal's behaviour can be interpreted more effectively than if only behaviour was monitored.

4.1   Health

Laboratory dogs often live in strictly controlled environmental conditions. These dogs are valuable and it is important to establish health programmes to ensure they are routinely treated for parasites, both external and internal, and vaccinated against all the relevant infectious diseases. Sick and injured animals should be identified and treated appropriately. Health programmes can be easily developed both in facilities where dogs are produced for laboratory use, and later in laboratories where dogs may come from breeding facilities or other sources. The latter may need to be quarantined for some time to ensure they do not infect the dogs already in the laboratory.

4.2   Nutrition

Dogs in laboratories should be fed a high-quality food, in accordance to their requirements, and be maintained at a correct body condition score. Dog-producing facilities should feed brood bitches, pups and juvenile dogs as appropriate and thus optimise growth, reproduction and rearing success.

4.3   Training

The techniques used to train dogs to be easy to handle and use in research are common to all training programmes. Positive reinforcement is the fundamental technique and staff engaged in rearing and training pups should be familiar with the theory and practice of operant and classical conditioning as they pertain to training dogs. When they are 4–14 weeks of age, pups should be socialised to humans and other dogs, and exposed to a variety of environmental factors which they will encounter in the laboratory. They should become accustomed to being on tables, to hoses, bathing, different people, outdoor cages and grass for elimination and exercise if necessary.

Laboratory dogs need to be confident in the laboratory without being too boisterous and active, and need to be trained to be quiet and relaxed. Obtaining the correct balance between confidence and over-activity is important to facilitate handling and research.

5.   HOUSING

The housing for any laboratory animal has to meet the requirements for hygiene, and ease of use, be economical in its use of space, and yet allow the animal to live in a stress-free environment and to behave in a reasonably normal manner and not exhibit abnormal behaviour. Animal housing must have an appropriate temperature (18–21°C), humidity (35–70%), change of air (8–12 per hour), and lighting (Bate, 1997). Dogs may be held in indoor pens with or without outdoor runs. Females in oestrus should be housed away from males. In many facilities, dogs are held alone in cages to minimise the spread of disease and reduce problems of aggression. Restricted physical and social conditions may result in abnormal behaviour but the influence of individual facets of the environment on such behaviour is poorly defined.

Dogs that do not cope with an inadequate housing environment may become chronically stressed, as indicated by physiological indices and behaviour (Beerda *et al.*, 2000). They may develop stereotypic locomotory behaviours such as circling, pacing and wall-bouncing (Hubrecht *et al.*, 1992; Hubrecht, 1995b. Hubrecht (2002) described one dog that engaged in such activity to the extent that it needed three times its daily food ration. This dog would seriously compromise any research project, but dogs with a less obvious problem may also likely influence the validity of results. Stereotypic behaviours usually stop when handlers enter the rooms and thus they may be seriously underestimated and require remote monitoring for them to be identified (Hubrecht, 2002).

Many dogs are used for long-term research, and it is important that they live in hygienic conditions to reduce the likelihood of infection and disease. Traditionally, animal houses are cleaned every day, and grated flooring that allows urine and faeces to be easily hosed away are popular. However trials have shown that dogs prefer solid flooring, and the use of small amounts of sawdust to soak up urine makes it possible to reduce wet cleaning of pens to once-weekly or even less if cleaning without water is used to remove faeces and soiled sawdust (Hubrecht, 2002).

Dogs have been companion and working animals for millennia but we have a poor understanding of their requirements for space. This is due to the great variety in types and sizes of dogs, and the wide range of systems under which they have been kept. However, it is important that when we keep dogs in cages, as in a laboratory, that the size and design of the cage be appropriate for the number and type of dogs housed in them.

The facility design is important for the welfare of individual and groups of dogs. The dimensions must allow for cleaning, the provision of enrichment devices, and ease of human contact and interaction. Cages high enough for the dog to stand on its hindlegs without touching the roof will not necessarily allow a human to enter and interact easily with the dogs if they require environmental enrichment or medical attention. However, dogs use the facility much more than humans do and it must be designed to meet the dogs' requirements.

The space requirement of a laboratory dog such as a beagle can be determined scientifically by monitoring the behaviour of individual dogs in cages of different dimensions (Beerda *et al.*, 1999a). This can be achieved by determining how much effort a dog will expend in order to gain access to cages of different dimensions, and by examining the effect of living in cages of different dimensions on their physiological and immunological responses (Beerda *et al.*, 1999 a, b) and the health of individual dogs. The behaviour of dogs, both as individuals and in groups, in cages of different dimensions has been observed by several groups of researchers (Neamand *et al.*, 1975; Hite *et al.*, 1977; Pettijohn *et al.*, 1980; Campbell *et al.*, 1988; Hubrecht *et al.*, 1992; Bebak & Beck, 1993).

When the size of a cage was small ($0.5–3.0$ m^2) and increases in its size were also small ($1–1.6$ m^2) there was little effect on the exercise behaviour of the dog. In addition there was no difference in the behaviour of dogs housed individually in cages 4.13 m^2 or 6.83 m^2 in size (Hubrecht *et al.*, 1992). This suggests that either the dimensions of cages are of little consequence to the behaviour of dogs, or that the size increases were not great enough to modify behaviour. Indeed, in a social animal, such as the dog, many other aspects of the environment might influence its behaviour. Regardless, dogs in these cages spent most of the day inactive (Hubrecht *et al.*, 1992).

However, there are distinct advantages in having larger cages for dogs. Deep cages allow dogs to avoid unpleasant experiences at the front of the cage, provide space for a kennel, allow enrichment objects to be placed in the cages, and let the dog to have a toilet area away from the feeding place and bed. Larger cages allow for more complex environments to be developed and permit dogs to be housed together, a feature which is probably more important for the dog's well-being than space. In the UK, minimum dimensions for laboratory dog cages are probably a compromise between financial and management requirements, and canine needs (Table 2).

The provision of a kennel in a communal cage allows the dog to avoid its cage mate and have a quiet place to live away from other dogs. Hubrecht *et al.* (1992) found that laboratory dogs spent 35% of their time in a kennel and he considered the kennel to be an important resource for dogs living in groups of five animals in pens 3.66 x 1.83 m in dimension, as it allowed them choice as to where they spent their time. A kennel for dog housed alone also gives them choice as to where to spend time and the opportunity to get out of sight.

Table 2. Minimum space recommendations for laboratory dogs of different body weights.

Bodyweight	Minimum floor area (m^2) per dog		Minimum height
(kg)	Housed singly	Housed in groups	(cm)
UK Home Office (1989)			
<5	4.5	1.0	150
5–10	4.5	1.9	150
10–25	4.5	2.25	200
25–35	6.5	3.25	200
>35	8.0	4.0	200
Canadian Council on Animal Care (1984)			
12	0.75	1.5	80
15	1.2	2.0	90
Instituted of Laboratory Animal Resources (1996)			
15	0.75		
Up to 30	1.08		
30	2.16		
MacArthur (1987)			
15–30	**4.2 (may be used for two animals)**		

6.   SOCIAL ENRICHMENT

Dogs are social animals and housing them in isolation is undesirable on behavioural grounds. However, it may be necessary to house dogs individually because of particular experimental protocols, the danger of aggression, during quarantine, post-operative convalescence, or the presence of exposed indwelling catheters.

In laboratories, it is possible to keep dogs in groups if the facilities are designed to allow this. It is obvious that the larger the size of the group the more complex the social life of each animal and the more socially-enriched their environment. However, there are difficulties and dangers regarding large groups. It may be difficult to move animals into or back into large groups because of aggression. This is a particular problem for dogs returning after being used for research which may debilitate them. They may become victims of aggression, especially if they had been high up the hierarchy previously. As aggression may be a problem with dogs housed in groups it is important that compatible animals are housed together and that groups of dogs remain stable. Moving animals out of and into groups may disturb the hierarchies which have developed, and result in aggression. Groups are best established when the dogs are young, but if new animals have to be introduced then it may be best to do so, on neutral ground, such as a new cage, rather than in the cage where the group has lived previously.

Large groups may be difficult to manage during research and dogs may have to be placed in individual cages, small groups or pairs. This may be disruptive and cause problems. Managerially, the smaller the group the better, and as Hubrecht (2002) found that there were no clear data indicating what the optimum group size might be, housing dogs in pairs seems a reasonable compromise. Regardless, dogs in pairs spent a similar proportion of time interacting with each other as did dogs in large groups (Hubrecht, 1993). However, the pairing of dogs may be insufficient to prevent chronic stress developing (Beerda et al., 2000) and dogs may actually need to live in larger groups even if other elements of environmental enrichment are possible.

Housing dogs individually resulted in abnormal behaviour (Hetts et al., 1992; Hubrecht, 1995b) and allowing them to live in pairs or larger groups is almost certainly an improvement. Dogs housed individually were more likely to engage in repetitive behaviours (Hubrecht et al., 1992) than those housed in groups. These authors speculated that dogs housed individually spent a lot of time trying to increase sensory input and spent much time standing on their hindlegs looking at neighbouring pens and the door. The presence of an additional cage mate allows for social interaction and makes the olfactory environment much more interesting. Physical isolation has deleterious effects on the behaviour of dogs, and allowing dogs to see other dogs might also improve their welfare. Visual contact with other dogs may stimulate allelomimetic behaviours such as barking, but may also make life more interesting and encourage normal behaviour (Wells & Hepper, 1998).

Dogs housed in groups spent more time walking and less time resting, and they exhibited less repetitive behaviour than those housed singly (Hubrecht et al., 1992). Housing dogs in groups allows for complex social interactions, but also guarantees a larger size of pen which may result in greater opportunity for the energetic trotting and running noticed by Hubrecht *et al.* (1992). When 12 beagles, normally caged in pairs, were allowed to mix for one hour each weekday they appeared to enjoy the interaction, but some male dogs developed skin and coat problems and there was an increase in the chewing of cage furniture. Hubrecht (1993) concluded that this method of social enrichment was of marginal value. However, the effect of allowing dogs, usually caged as individuals, to mix with other dogs needs further investigation.

Pairs of compatible dogs can be used in most types of research, including toxicological and nutrition trials. The dogs can be separated for dosing and feeding but otherwise held together. Dogs may have to be held as individuals following surgery, but even then they should be allowed to have visual contact with other dogs.

7.   HUMAN-DOG INTERACTION

Dogs from well-run breeding facilities will be socialised with humans during the first few months of life and will find interaction with humans a good source of social complexity. Dogs that were given extra human contact as pups were more approachable at 6 and 11 months of age and appeared to seek human contact (Hubrecht, 1995b). Grooming by humans of dogs for 30 seconds each day reduced cage furniture-chewing by 90%, and the dogs became more approachable towards their regular handlers and strangers (Hubrecht, 1993). Regular 90-minute walks for dogs housed individually may significantly reduce the chronic stress experienced by such dogs (Beerda *et al.*, 2000), but this may be due to the varied and interesting environment the dog experiences during a walk rather than human contact.

Contact with humans may be extremely important for dogs and some believe it may be more important than contact with other dogs (Wolfle, 1987, 1990; Fox, 1986; cited by Hubrecht, 1995). Handling by humans may reduce blood pressure in dogs (Lynch & Gantt, 1968) and reduce their physiological response to various stressors (Tuber *et al.*, 1995; Hennessy *et al.*, 1998). However, the exact nature of the interaction appeared important, and under some conditions petting by men increased plasma cortisol concentrations of dogs (Hennessy *et al.*, 1997). Gentle, soothing speech seemed to be important in reducing or preventing a hypothalamus-pituitary-adrenal axis response to venepuncture (Hennessy *et al.*, 1998). Thus, handling in itself may not be effective in reducing the stress response, and specific types of handling may be required for dogs reared under different conditions.

It is probably better for laboratory dogs if the people responsible for managing them in the research facility have knowledge of dogs outside the facility. If the people handling research dogs have pet dogs at home they may be more sympathetic to the research animal and be quicker to identify when it has a problem. The attitude of people towards the animals in their charge is important in that animals treated improperly may develop chronic fear (Hemsworth & Coleman, 1998).

The impact of the relationship between experimental animals and humans on the physiology and behaviour of the experimental animals may impact on the quality of the research (Gross & Siegel, 1979; David & Balfour, 1992). This has not been assessed in dogs but its impact may be even more significant in dogs than other species, given the close relationship between humans and dogs.

8.   ENVIRONMENTAL ENRICHMENT DEVICES

A variety of enrichment devices have been trialled with pet and laboratory dogs, and have been found to modify behaviour and reduce abnormal behaviour. Enriching devices need to be safe, tough and suited to the design of the cage. The biological significance of these devices remains unclear, as simply changing behaviour may not in itself be significant with regard to the biological fitness of the animals concerned. However, if more complex behaviour results from the use of such devices then this is generally taken to be a positive outcome. The provision of toys (chew bone, rawhide, plastic piping) modified the behaviour of caged beagles by decreasing the time spent inactive, but dogs with toys spent less time in social interaction and toys had little effect on the development of stereotypic behaviours (Hubrecht, 1993). Whether this change in behaviour reflects an improvement in welfare is more difficult to assess. The toys were chewing toys and this might have resulted in more covert competition between dogs. It certainly reduced the chewing of cage furniture.

When a set of steps was introduced into a cage it increased the complexity of the pen and added a vertical component. Dogs spent time on top of the steps, on watch, and also time under it hiding. If the provision of such devices does not reduce the incidence of abnormal behaviour, particularly stereotypic behaviours, then they may not have much of an impact on a dog's welfare. More detailed studies of the effect of such devices on physiological and immunological responses are required, as whilst changing and allowing for more complex behaviours may be welcome it may be of little biological significance. Steps or a kennel provide dogs with a choice and some control over where they spend their time. The provision of such simple options is generally considered important in improving the welfare of caged and penned dogs. Other useful enrichment tools include odours (Wells, 2004), and the use of television or video-projected images should be investigated, as dogs will pay attention to moving images in various situations (Pongracz *et al.*, 2003).

9.   NOISE

Kennels are notoriously noisy, and most of the noise originates from dogs barking but other sources of noise include doors banging, equipment, and human conversation and laughter. In addition, noise from outside the facility, such as traffic, may affect the animals. Dogs hear a wider range of sounds than humans, from 0.04 kHz to 50 kHz, and also hear at much lower levels than humans. In laboratories, sound often reached high levels of over 85 decibels and up to 125 decibels (Sales *et al.*, 1997), which can cause stress to humans, and presumably also to dogs as they are more sensitive to noise than humans (Ottewill, 1968; Petersen, 1980). Much of the noise may be made by dogs barking. A high level of barking may damage the hearing of dogs, but, this has not been clarified. The noise in a laboratory will vary during the day, decreasing at night and increasing towards morning (Sales *et al.*, 1997). Therefore, it is important to locate and design laboratories so as to minimise noise by using sound-depressing materials, such as cavity walls, and sound-reducing doors. Individual rooms connected by corridors may be better than large open facilities. Dogs can also be trained to be quiet during periods known to stimulate barking, such as before feeding.

Appropriate music for dogs and humans may improve conditions in laboratory dog houses. Classical music apparently relaxed dogs in animal shelters while heavy metal music increased barking, but other pop music had no effect (Wells *et al.*, 2002). Music may dampen out other noises and make the environment more stable.

10.   EXERCISE

Exercising kennelled dogs intuitively appears to be a worthwhile procedure for improving the dog's welfare, and this has led to the Congress in the USA stipulating that minimal standards of exercise for dogs be established (Clark *et al.*, 1997). Increasing the size of a dog's kennel did not appear to increase the dog's level of fitness (Clark *et al.*, 1991). In an attempt to evaluate the effect of exercise, Clark *et al.* (1997) monitored behaviour, immune function and plasma cortisol concentrations in beagles that were caged singly and either exercised alone, with a con-specific or not exercised at all. The exercise comprised being placed in a room for 20 minutes thrice-weekly for 12 weeks. This activity no significant effect on the physical health of the dogs but then placing a dog in a barren room was unlikely to stimulate much exercise in any case.

More extensive exercise, such as allowing dogs to exercise outdoors either with other dogs or a human, has been recommended as essential for the mental and physical health of dogs, but this has not been proven (Loveridge, 1994; Trussel et al., 1999). Whilst outdoor exercise is attractive from the dog's perspective, the fear which staff in laboratories using dogs have, of animal rights activists may make it difficult for such laboratories to have outdoor exercise yards for their dogs. Another possible problem is that some research institutions may move from countries with a populace sympathetic towards dogs to countries with less sympathetic attitudes towards them.

The design of dog facilities should allow for ease of husbandry, aim to reduce noise, allow dogs to live in pairs or groups and to see other dogs, and allow for as wide a variety of behaviours as possible. Dogs housed individually or in pairs for years and not walked frequently will experience chronic stress (Beerda et al., 2000) and to reduce such stress housing for dogs needs to be much more dog-orientated and allow easy dog and human contact (Loveridge, 1998). The design of facilities should allow for pens to be joined so that large pens can be created. This might involve placing small pop-holes between pens (Hubrecht, 2002).

Regular walking and play are recommended. Dogs experiencing chronic stress are not good models for biomedical research of any kind, and housing and management must either adapt to produce dogs that are not stressed or not use them in research at all. In Australia, the National Health and Medical Research Council (1996) recommends that dogs be taken outside to run freely or walk on a leash for at least 30 minutes each day, even in bad weather.

11.   RE-HOMING LABORATORY DOGS

Some dogs spend a short period of time in laboratories where canine research is infrequent and not ongoing. Dogs for this type of research may be obtained from breeding facilities or purchased elsewhere. If the former, they are probably not suited for re-homing as they are institutionalised and might not easily adapt to household living and may have particular behavioural problems, such as being difficult to house train or barking incessantly. They may also miss the social life they had previously.

Dogs obtained from pounds or shelters may return to where they came from and be re-homed from there, if considered appropriate. Their experience in a short-term experiment may actually make them more suited for re-homing, if they have been trained and managed effectively and sympathetically.

The majority of laboratory dogs are probably humanely killed rather than being re-homed, and so it is extremely important to give them as rich a life as possible as it will virtually all be spent behind bars.

12.   SURGICAL RESEARCH AND ANALGESIA

The use of any animals, but especially the dog in biomedical research, is an anathema to many people. These animals have been our companions and helpers for thousands of years and the argument is that it is immoral to use them for research purposes. In most countries, however, it is legal to use dogs in research and whilst this is unlikely to change in the near future, the legislation governing the use of dogs in research is often stricter than with other laboratory animals. In addition, laboratories and scientists using dogs are aware that they are particular targets for organisations against the use of animals for research and recognise that they have to manage and use dogs at the highest standards.

13.   RESEARCH AT HOME

Much of the research targeted at the health, behaviour and welfare of companion dogs uses dogs owned and living at home. Although this latter type of research is carried out with the consent of the dog's owner, it needs to be controlled so as to ensure the owner understands what is happening and that the animals are not harmed by the research. This research needs ethical control similar to that used for medical research, as the owners are equivalent to the parents of children that may be involved in clinical or psychological research.

14.   CONCLUSIONS

The work of Beerda *et al.* (2000) suggests that many dogs held in laboratory cages are suffering form chronic stress, although caution was recommended in interpreting the data. This suggestion has serious implications for the quality of research work undertaken using such animals, and their welfare. In an ideal situation, laboratory dogs would be housed in groups, with indoor and outdoor runs, have plenty of appropriate human contact including walks outdoors, and be fed and managed appropriately. The ongoing keeping of laboratory dogs indoors in individual cages may be bad for science and will continue to attract criticism from animal welfare and rights organisations. However, facilities such as those described by above (Loveridge, 1998) may actually provide dogs with a better quality of life than many dogs living with their owners, at least for the time before they are used for invasive research.

Chapter 10

DOGS IN SHELTERS

Abstract: There are unwanted dogs in all countries and their fate is a reflection of the
              wealth of these countries and people's attitudes towards animals and dogs in
              particular. In most countries, there are animal shelters of some description
              which take in unwanted dogs and either re-home or kill them. The number of
              unwanted dogs and dogs entering shelters is decreasing in many European
              countries, the USA and Japan, and possibly worldwide. The time dogs spend
              in shelters and the quality of life therein impacts directly on their welfare and
              may affect their subsequent behaviour if they are re-homed. Dogs are
              surrendered to shelters for various reasons, many are unwanted puppies. While
              shelters attempt to maximise the welfare of their inmates, many dogs are held
              alone in single cages and may not receive sufficient attention due to shortages
              of staff and volunteers.

1.   INTRODUCTION

There are unwanted dogs in most countries and these dogs are either
ignored, killed, or or they may be captured and rehomed. Dogs that are free-
ranging, whether owned or not owned, may be caught and put in a pound or
shelter. If owned but unwanted, they may be surrendered directly by their
owner to a shelter. Shelters managed by animal welfare agencies and pounds
managed by local authorities have different philosophies with regard to how
to manage their inmates. The majority will hold dogs for a limited period of
time, usually about a week, return those claimed by their owner, attempt to
re-home those considered suitable for re-homing, and then kill the
remainder. Some shelters have no-kill policies and will hold dogs until they
are reclaimed, re-homed, are humanely killed for medical reasons, or die.

179

Pounds operated by local authorities may not re-home dogs and only return those that are claimed by their owner and kill the rest.

Animal shelters exist worldwide, and for the last 100 years or more the animal welfare organisations that manage them have worked to educate the public about the unacceptability of allowing unwanted pups to be born and of the responsibilities of dog ownership. Although many claim that the problem with unwanted dogs is growing, the number of animals sheltered by the American Society for the Prevention of Cruelty to Animals (ASPCA) in New York and the proportion killed has decreased and the percentage re-homed has increased over the last century (Table 1) (Zawistowski et al., 1998). Also in New York, the number of dogs killed in relation to the human population has decreased. In Oregon USA, a similar pattern was found and the number of dogs in a shelter decreased from 28,850 in 1973 to 6,665 in 1991, and the number killed decreased from 11,566 to 1,171. This was accompanied by a decrease in numbers of dogs in a county pound from 16,000 to 6,000 over a similar period (Strand, 1993). The American Humane Society reported a decrease by 40% in the number of animals sheltered from 1985 through to 1990 (Strand, 1993). This occurred despite the number of dogs increasing in the USA over the last few decades. A similar pattern was noted in Japan (Table 2) (Hart et al., 1998), and the downward trend in the population of dogs in some European countries will almost certainly be paralleled by a decrease in dogs entering shelters and pounds. Apparently, some European shelters are now importing dogs from other countries to fill their requirement for dogs of re-homing, and possibly to justify their existence.

Table 1. American Society for the Prevention of Cruelty to Animals figures for dogs received, re-homed, or humanely killed over much of the last century in New York.

Year	Dogs received	% Dogs adopted	% Dogs killed
1895	21,741	1.8	95.7
1896	27,587	2.4	86.6
1904	30,505	Na	93.7
1914	59,355	Na	88.0
1928	85,744	3.2	82.3
1934	65,207	6.1	93.5
1946	60,537	2.9	91.0
1954	66,043	13.6	73.7
1965	70,185	21	65.1
1974	81,627	8.8	82.7
1984	41,867	19.4	75.0

na = not available
Adapted from Zawistowski et al. (1998)

Dogs become unwanted because some people find they cannot manage them effectively. The desire to have dogs may be encouraged by personal experience of the pleasures of having a dog, belief that a dog is good for one's health and family, desire for a working or guard dog, powerful advertisement to promote the benefits of pet ownership, and many other reasons. The presence of unwanted dogs has been blamed on irresponsible owners who will not de-sex their animals, ignorant owners who believe that it is good for a bitch to have one litter, breeders, veterinarians who charge too much for de-sexing, the pet food industry, and pounds and shelters (Sturla, 1993). These accusations have sometimes resulted in poor cooperation between interested parties (MacKay, 1993).

In developed countries, the problem is basically due to individual owners (Olson and Moulton, 1993) breeding bitches and producing unwanted pups. This may be due to ignorance, laziness or poverty, but it cannot be blamed on the many organisations and professionals who have encouraged responsible ownership for decades and now provide more service, both in training and private veterinary care, than previously. A small proportion of the dog-owning community is probably to blame for surplus pups and this population needs to be targeted if the problem is to be reduced further (Murray, 1993).

Table 2. The population of dogs (millions) and the numbers impounded, killed or given to laboratories, reclaimed or adopted, in 1984 and 1994 (Hart *et al.*, 1998) and 2000 in Japan.

| | | Fate | | |
Year	Impounded	Reclaimed	Adopted	Killed (laboratory)
1984	345,136	13,714	NA	331,422[a]
1994	243,753	13,131	10,791	197,789 (22,042)
2000	126,570	15,004[b]	111,566	

[a] Killed or given to laboratories

[b] Reclaimed and adopted

NA = not available

2.   HISTORY OF SHELTERS

The Society for the Prevention of Cruelty to Animals (SPCA) was established in England in 1824. Initially it was primarily concerned with the welfare of horses but it gradually came to recognise that the welfare of dogs was also a major issue and dog shelters were established. The British Empire was in its heyday and the philosophy of the SPCA and the concept of animal shelters spread thorough the empire. In the USA, the ASPCA was established in 1866 by Henry Bergh, and again horse welfare was central to its activities.

In the 19th Century in many cities, dog control officers attempted to remove stray dogs from the streets to reduce the likelihood of rabies and attacks on humans. These dogs, if unclaimed, were killed often in quite brutal fashion by clubbing or drowning. The SPCA became concerned about the fate of these dogs and in some cities, such as New York, the SPCA took over dog control from the city authorities (Zawistowski *et al.*, 1998). In many cities, however, municipal pounds and SPCA shelters existed side-by-side and do so to this day worldwide.

Traditional animal welfare organisations such as the SPCA have established and run animal shelters that generally take in unwanted dogs surrendered by their owners, stray unowned dogs at risk of public abuse, injured dogs, and unwanted pups. These shelters either re-home these dogs or humanely kill them if they are unwanted or unsuitable for re-homing. These organisations have worked for decades at trying to increase responsible ownership, and while there have certainly been improvements there are still too many pups produced, and many dogs still become unwanted, and free-ranging dogs remain a nuisance (Murray, 1993). It is important to remember that most animal welfare organisations are charities and depend on public donations and legacies to survive. Thus, there is a tendency amongst such organisations not to trumpet success as it might result in lower donations for future work.

In North America and Europe, there is a plethora of private non-profit-making shelter and animal control agencies that take in free-ranging and unwanted dogs. In Japan, homeless and unwanted dogs are managed almost exclusively by government animal shelters (Hart *et al.*, 1998). Some animal welfare organisations have 'no-kill shelters' but these are in the minority as most charities cannot afford to keep dogs that are not suited for re-homing. Dogs may be considered unsuited for re-homing because of their behaviour or because they are of a specific breed or crossbreed. In addition, some veterinarians now refuse to kill healthy animals and these may also be surrendered to shelters. In some religions, killing animals is not allowed or is disapproved of and this compounds the difficulties of managing free-ranging and unwanted dogs.

Dog pounds are built by local authorities and managed either directly or by private companies or welfare organisations. The function of a pound is to provide a place where stray and unwanted dogs can be held until they are either returned to their owner, killed or re-homed. The rules governing each pound will differ depending on local by-laws and the philosophy of the staff. The by-laws will be influenced by national dog control laws, animal welfare legislation and other pertinent legislation.

3.   HOW LARGE IS THE PROBLEM?

The number of organisations involved in animal shelter work and the number of shelters make it difficult to determine accurately the number of animals entering shelters and their fate (Zawistowski *et al.*, 1998). There are about 5,000 shelters/and pounds in the USA, each housing more than 100 dogs per year, but to garner data from them is difficult. Over two million (2,112,009) dogs entered the 1,100 shelters surveyed by Zawistowski *et al.* (1998) in 1994. About half of the dogs (51%) were admitted by animal control personnel and 27.7% were relinquished by their owners. Of these dogs, 25.4% were adopted, 15.6% reclaimed by their owners and 56% (about 1.1 million) were killed. It is impossible to know if these shelters are representative of others, but these figures suggest that about five 5 million dogs are killed in shelters and pounds each year in the USA.

The estimated population of dogs in the USA varies from 44 to 55 million, depending on the methodology used (Patronek & Rowan, 1995). These figures are based on household ownership of dogs and probably do not include many of the unwanted pups born and surrendered to shelters early in life, and free-ranging dogs. In 1969, Marx and Furculow (1969) reported that 20% of the population of dogs in the USA was feral, but this has declined significantly and Patronek and Rowan (1995) considered the number of feral dogs in the USA to be insignificant. Thus, approximately 20% of dogs in the USA enter a shelter each year and half of these are killed. These figures are higher than those of Patronek and Rowan (1995) who estimated that each year, about four million dogs enter shelters and 2.4 million of them are killed. The figures quoted by Alexander and Shane (1994) that between 12 and 20 million animals are killed each year in animal control facilities seem extreme, despite the inclusion of cats and other species. Lower figures are supported by findings that in 1996 in Ohio (Lord *et al.*, 1998), which has a population of 2.2 million dogs, over 200,000 (about 10%) dogs entered a shelter and over 128,600 were killed. Similar figures are quoted by Arkow (1994), who found that in California, Colorado and Iowa about 9.25% of dogs were assisted by animal control personnel annually.

In Japan (Table 2), less than 2% of dogs entered a shelter and the majority were killed (Hart *et al.*, 1998). In the UK, about 100,000 dogs were re-homed by the SPCA in 2003, when the population of dogs was about 6.2 million, that is, less than 2% of dogs. In New Zealand, with a population of about 500,000 dogs, the SPCA received about 10,000 dogs in 2000 (that is 2% of the dog population) (Table 3) and about 45% (4,500) of these were killed (Peter Blomkamp, RNZSPCA, New Zealand, personal communication). In addition, municipal dog pounds probably take in at least the same number of dogs and kill the majority of them, thus adding probably another 10,000 dogs to the numbers above. This would total about 4% of the total dog population being killed in shelters and pounds, a number significantly lower than the USA. In the Australian cities of Melbourne and Brisbane, about 10% of the dog population was killed in shelters and pounds (Murray, 1993). However, in Australia, as in Japan, the number of dogs entering shelters is decreasing as is the number of dogs being killed (Table 4). None of the figures shown above includes the number of unwanted dogs killed by veterinarians in private practice.

Table 3. The population of humans (millions) and dogs (millions), and number of dogs (millions) entering shelters each year in Japan, USA and New Zealand.

Country	Human	Dogs	Shelters
Japan	127	4.1	0.243
USA	296	50	4
New Zealand	4.1	0.5	0.02

Lepper *et al.* (2002) identified the characteristics of dogs adopted from a shelter in California and suggested that it would be worthwhile promoting certain other types of dogs for adoption. However, in that particular shelter 26% of the 7,720 dogs impounded were put up for adoption and only 16% of all the dogs were adopted.

Dogs are a nuisance in many societies. In an Australian study, Murray (1992) found that dogs were the second largest problem for municipal management, after the collection of rates, and in Perth 20% of dogs were allowed to roam freely (Adams & Clark, 1989). There are different perspectives on the problem of unwanted dogs. In an Australian provincial city, veterinarians and shelters killed the same number of dogs but the shelters killed a greater proportion of young unwanted animals than the veterinarians. Shelter workers saw over-breeding as a major problem, but this was not obvious to the veterinarians in their day-to-day work (Murray & Speare, 1995). This observation may be true for private veterinarians worldwide, who kill numerous dogs but these are of a different demography than those killed in shelters.

The problem of unwanted dogs is probably decreasing in wealthy countries, but it remains a problem, and is a much greater problem in many poorer countries, where there are insufficient finances to cope. In poorer countries public health initiatives to reduce the number of unowned dogs are usually short-term population-reducing programmes. Long-term support for public education programmes and the establishment and support of shelters in these countries are a challenge for dog lovers in the wealthy world.

4.   DOG POPULATION DYNAMICS

A population model developed by Patronek and Rowan (1995), based on dogs in the USA, used a birth and death rate of 12% and a turnover rate of 14%, which included stray dogs returning to the population of owned dogs. Of the dogs that died in this model (6.2 million), over a third (2.4 million) died in shelters. The other 3.8 million dogs died in veterinary hospitals, at home and at large. The figure of 12% appears to be acceptable for the birth and death of dogs in the population of owned dogs in developing countries but not in poorer countries (see Chapter 2). In 1990, Arkow (cited by Kidd *et al.*, 1992a) stated that the average stay of a pet in an American home was about 2 years, which would suggest a much greater turnover than the 14% proposed above.

Table 4. Australian SPCA figures for dogs entering shelters, re-homed or killed humanely.

	Year	
Fate of dogs	2002–3	1998–9
Received	64,593	72,360
Reclaimed	14,788 (22%)	
Re-homed	21,469 (32%)	38,464[a]
Killed	24,554 (38%)	33,896

[a] Reclaimed and re-homed

5.   WHY ARE DOGS SURRENDERED TO ANIMAL SHELTERS?

There are four major reasons why a dog enters a shelter or pound, *viz* caught as a free-running animal, surrendered by their owner, found injured in a public place, or seized as part of an animal welfare investigation. Although the reason for entering a shelter/pound varies between countries, and probably over time, the majority are caught as free-running animals (Table 5). Many of the animals surrendered are from unwanted litters of pups either given up by the owner or collected in a public place. In individual shelters, the reasons for entering probably remain stable over the years (Alexander & Shane, 1994). In Las Vegas, USA, dogs acquired from breeders and pet shops were less likely to be surrendered to a shelter than dogs acquired elsewhere (Nasser *et al.*, 1984), and Arkow (1985) stated that typically an unwanted dog was acquired at no or low cost and for compassionate reasons rather than for a specific activity.

People relinquish their dogs for a myriad of reasons, which includes personal circumstances, the dog's behaviour, or simply that the dog is unwanted (Table 6). In a large American study of 12 shelters in four regions, the background to why 3,676 dogs and litters were surrendered was investigated (Salman *et al.*, 1998; New *et al.*, 1999; Scarlett *et al.*, 1999; Kogan *et al.*, 2000). The reasons, in decreasing order of importance for surrendering dogs were, human housing issues, behaviour, human lifestyle problems, requests for euthanasia, human expectations and preparation, household animal population, animal health, and animal characteristics (Salman *et al.*, 1998).

Table 5. The reason why dogs entered animal shelters or pounds, in different countries.

	Japan[1]	New Zealand[2]	USA[3]	USA[4]	Northern Ireland[5]	Australia[6]
Number	6,884	967	2,112,009	214143	18,843	20,729
Year	1994	1999–2001	1994	1996	1990	2001–2
Caught as stray	61%	30%[a]	51%	66%	54%	84%
Relinquished	22%	63%	27.7%	32%	27%	15%
Cruelty				1%		
Injured	4%					
For euthanasia					19%	
Other	13%	7%	22%	1%		1%

[a] Data from SPCA (municipal pounds usually take in stray dogs)
[1] Hart *et al.* (1998)
[2] Phipps (2003)
[3] Zawistowski *et al.* (1998)
[4] Lord *et al.* (1998)
[5] Wells and Hepper (1992)
[6] Marston *et al.* (2004)

Scarlett *et al.* (1999) analysed the reasons why people from 520 households relinquished 554 dogs (Table 6). These reasons, in descending order of importance were; no time for pet, personal problems, allergies, pet-child conflict, and a new baby, and also included divorce, travelling, unwanted gift, and owner deceased. Of interest is the observation that dogs came into and went out of more than 50% of those households during the previous year, suggesting poor stability in the human-animal relationships. More than half (59%) of these dogs were acquired at no cost. Males and females were equally likely to surrender a dog, but people 25–39 years of age were over-represented among those who surrendered a dog because of moving house (New *et al.*, 1999).

In that large study, equal numbers of male and female dogs were surrendered, about 43% of which were neutered, 68% were mixed breed, and 27% were 'outside' dogs. The dogs were sourced from, in descending order of importance, friends, shelters, breeders, strangers, or found as strays, and 46% had been owned for less than a year (Salman *et al.*, 1998). When dogs were surrendered for a behavioural problem, the problems, in descending order of occurrence were; bites, aggressive towards people, escapes, destructive inside, destructive outside, disobedient, problem between new pet and other pets, aggressive towards animals, soils house, and vocalises too much (Salman *et al.*, 2000). One half of dogs surrendered for a behavioural reason had been acquired from a shelter, most lived with the owner for less than 3 months, 58% lived in households with another dog or cat, and for 51% another dog had been added to the household in the previous year. Living in a multi-animal household with regular additions increased the chances of being surrendered for behavioural reasons, and there was a short period of time to deal with any problems that developed (Salman *et al.*, 2000). In the Netherlands, aggression and separation anxiety were the most common reasons why dogs were surrendered or returned to animal shelters (van der Borg *et al.*, 1991).

It is easy to think badly of people who surrender their dog to a shelter and that their decisions to do so are trivial (DiGiacomo *et al.*, 1998), but this is not the case for many. In a series of interviews with people who had relinquished dogs, those authors found that all had delayed surrendering their animal for some time as they balanced their concepts of what would happen at the shelter with their circumstances. Procrastination, and failed attempts at resolving the problem made by poorly-informed owners were common observations. The authors concluded that most relinquishment cases were more complex than the recording of one or two reasons would imply. Blame displacement mechanisms were used by many owners to cope with the possibility that the surrendered animal would be killed (Frommer & Arluke, 1999).

People bringing animals to shelters for re-homing or euthanasia are doing what shelters would recommend. However, many of the reasons why animals are relinquished could be either prevented or treated with proper advice, and perhaps people should additionally regard shelters as a source of information and advice (DiGiacomo *et al.*, 1998). Indeed, Murray (1993) suggested that animal shelters may encourage irresponsible pet ownership by providing a convenient place for people to obtain pets and then to relinquish them when it suited.

The major risk factors associated with being relinquished to a shelter for a dog (Patronek *et al.*, 1996) were; being sexually intact, obtained at no cost (but not as a gift), older at acquisition, being young, spending much time in a crate or yard, owners recognising that caring for the dog was harder work than they had thought, lack of veterinary care, lack of attendance at training classes, and inappropriate elimination. Reaching new and at-risk owners and helping them reduce the likelihood of surrendering their dog is important. Increasing the percentage of animals de-sexed, encouraging veterinary care and advice, and attendance at training classes are all important. The majority of surplus dogs are usually unwanted pups produced by uncontrolled mating, and this remains a major problem worldwide. In many cities, there are increasing opportunities to attend dog training classes and there are more veterinarians involved in private companion animal health practice, so these risk factors should become less of a problem. However, underlying them may be the reality that caring for a dog takes time and costs money, and both these factors can only be appreciated through constant education and advice from everyone involved with dogs.

Table 6. The reasons why people relinquished their dogs to a shelter.

	USA[1]	USA[2]	USA[3]	NZ[4]
Number of dogs in study	269	53	2,045	967
Unwanted litter	36%	5%		25.5%
Unwanted animal	39%			13.9%
Animal ill		4%	8%	
Owner moving/landlord	8%	19%	30%	8.9%
Financial problems	2%	21%		7.9%
Euthanasia			18%	
Inappropriate behaviour	9%	30%	47%	6.5%
Owner ill		9%	27%[a]	
Other	7%	12%	5%[b]	6.9%
			14%[c]	
			7%[d]	
Stray				30.4%

[a] Human health and personal issues
[b] Animal characteristics, not medical or behavioural
[c] Human preparation or expectations
[d] Household pet population
[1] Alexander and Shane (1994)
[2] Miller et al. (1996)
[3] Scarlett et al. (1999)
[4] Phipps (2003)

6.   COPING WITH LIFE IN A SHELTER

The retention of dogs in shelters and pounds for long periods of time may have negative effects on their behaviour. On average, dogs were held for up to 85 days in a shelter in the Czech Republic (Nemcova & Novak, 2003), but for about two weeks in two Californian shelters (Clevenger & Kass, 2003). However, after five days of being held in barren cages, dogs appeared to get used to the facilities and ate their food quicker, and became relaxed in the company of a stranger, although their response to a novel object did not change. Responses in the cage were the same for stray (free-ranging) dogs and dogs surrendered by their owners (Wells & Hepper, 1992).

Many dogs in shelters are held alone in barren cages (Tuber *et al.*, 1999) in isolation, for quarantine and safety reasons. If they could see another dog through the gate of their own cage they spent time at the front of the cage, which may have improved their welfare (Wells & Hepper, 1998). There are no physiological data to support this, but it is intuitive. In Germany, about half of all shelters keep dogs in groups although there is a great fear of aggression. However, the majority of confrontations are settled by ritual without aggression, and dogs housed in groups had a closer relationship with humans than those housed alone (Mertens & Unshelm, 1996). Dogs that were housed individually were more likely to show stereotypes and other behavioural problems than those housed in groups. Behavioural problems are common in dogs adopted from (Voith & Borchelt, 1996) and surrendered to animal shelters, or caught as strays (Patronek *et al.*, 1996). More than half (53%) of the owners of dogs that had been housed in groups in a shelter were content four weeks after adoption, whereas 88% of those who had dogs that were housed individually complained of problems (Mertens & Unshelm, 1996).

The hypothalamus-pituitary-adrenal responses of dogs in shelters reflect their behaviour. Plasma cortisol levels were high during the first three days (Hennessy *et al.*, 1997) and declined to a baseline by Day 9 (Hennessy *et al.*, 2001), while levels during Days 4 to 9 were intermediate (Hennessy *et al.*, 1997). Levels during the first few days were nearly three times greater than for dogs in their own home, which suggests that that time in a shelter is stressful. This is not surprising, as it is known that exposure to novel or restrictive conditions elevates plasma cortisol levels (Tuber *et al.*, 1995). Interacting with a human for 20 minutes of petting did not affect cortisol levels (Hennessy *et al.*, 1998), but those dogs petted by a female had lower plasma cortisol levels than those petted by a male (Hennessy *et al.*, 1997). This effect was later shown to be due to the difference in petting technique, as in a follow-up study there was no difference when males and females petted dogs in a similar fashion (Hennessy *et al.*, 1998), although the behaviour of the dogs during petting was slightly different. The dogs yawned more and spent more time with their head up in a relaxed posture when petted by women than men. Thus, minor differences in the way dogs are handled appear to affect the HPA response, suggesting that dogs are particularly sensitive to how humans touch them. Dogs that pass through shelters are prone to separation anxiety (McCrave, 1991), and Hennessy *et al.* (1997) suggested that the experience might make dogs more sensitive to separation anxiety after re-homing. This may be due to the breakdown of the relationship with the previous owner and an overly close attachment to the new one (OFarrell, 1992; Askew, 1996).

Human interaction and basic training can make it easier for dogs to live in shelters and to accommodate the change to their new home. Tuber *et al.* (1999) described a programme to meet these goals. The programme included having the dog spend time each day in a 'living room' in the shelter, 20 minutes of human contact with firm gentle stroking, and training the dog to accept confinement in a cage, to sit when someone approached its cage, and on the approach of potential adopters, and to offer classes for new owners on how to train their dog.

Most animal shelter personnel consider that the animals' best interests are paramount, but there is little evidence in the literature as to how animals are actually managed in shelters, with regard to health, nutrition and housing. Because shelters continually take in dogs with unknown medical and vaccination histories, there is the constant danger of canine parvovirus disease, kennel cough and canine infectious hepatitis (Pratelli *et al.*, 2001). Internal parasites such as hookworm need constant veterinary care if they are to be controlled, especially in shelters where dogs are housed in groups (Kornas *et al.*, 2002; Svobodova, 2003), and in young dogs (Le Nobel *et al.*, 2004). In one survey of shelters and pounds in Ohio, USA, Lord *et al.* (1998) found that 45% dewormed and 43% vaccinated their animals, 22% undertook testing for heartworm, and 56% had a de-sexing policy.

Personnel working in shelters are often volunteers with limited knowledge of the training and management of dogs (Rusch, 1999) and this makes staff management difficult. In addition there may be a large turnover of personnel with the attendant difficulties of maintaining an educated and knowledgeable workforce. This may impact on the wellbeing of dogs in shelters. Staff working in shelters may be stressed due to working with unwanted dogs and being involved with euthanasia, which may impact on how they manage animals. Shelter staff reported less stress than those working with homeless humans (Ferrari *et al.*, 1999) and used blame displacement strategies to deal with their guilt concerning euthanasia (Frommer & Arluke, 1999)

7.   WHAT DETERMINES WHETHER A DOG IS RE-HOMED OR NOT?

The factors which determine whether a dog is put up for adoption or not vary depending on the philosophy of the shelter or pound and their management processes. Shelters which have a no-kill policy will attempt to re-home virtually all animals. Most shelters will identify those dogs that are easy to re-home because of looks, breed, size, age and behaviour; those that may be more difficult to re-home and may need some training or medical attention; and those that are difficult to re-home. Some shelters may have a policy which does not allow some dogs to be re-homed on principle or because they fear what might happen to the dog or its new owners. Pit bull terrier-type dogs are often not re-homed for these reasons (Lepper *et al.*, 2002). Dogs difficult to re-home might include breeds or types that the shelter knows it will have difficulty in re-homing, or very young pups that have not been with their mother for long or have not been socialised well.

Some research has been conducted over the past decade on the selection of dogs for re-homing. There is evidence that good pre-selection practices and post-re-homing education of the new owner may reduce the incidence of behavioural problems experienced by the new owners.

It is important to determine what factors make a dog attractive to a potential owner as limited space in most shelters may force them to kill dogs without putting them up for adoption (Posage *et al.*, 1998). Mixed-breed dogs were 1.8 times more likely to be killed than purebred dogs, and the chances of euthanasia increased with age (Patronek *et al.*, 1995). Breed did not make any difference in that study, but Posage *et al.* (1998) found that terrier, hound, toy and non-sporting breeds were more likely to be adopted. Lepper *et al.* (2002) found that breed was important to potential owners and purebred dogs were more likely to be adopted. Additionally, small size, a history of being indoor dogs and having a gold, grey or white coat colour were good predictors of being adopted. The likelihood of being adopted decreased with age. Brindle coloured dogs, and black dogs, were not popular,but red merle and tricolour dogs were.

The behaviour of a dog in its cage influenced whether or not it would be adopted (Wells & Hepper, 2000a). Dogs that were alert, quiet and at the front of the cage were more attractive to potential owners (Wells & Hepper, 1992, 2000a). Placing the dog's bed at the front of the pen, hanging a toy there, or increased human contact with the dog all increased the likelihood of a dog been adopted (Wells & Hepper, 2000a). Placing dogs in cages where they could see other dogs through the gate might also encourage them to stay at the front of the cage and thereby be more attractive (Wells & Hepper, 1998).

Dogs categorised as stray rather than unwanted were less attractive to potential owners (Wells & Hepper, 1992), and this was expected as stray dogs were more likely to show behavioural problems (especially a tendency to stray), when adopted than dogs which had been surrendered. However, Lepper *et al.* (2002) found that strays were more popular than dogs relinquished for behavioural reasons, or old and sick dogs, but less popular than dogs relinquished because of financial or home-moving reasons.

As all shelters fear re-homing dogs that then become dangerous to people, they attempt to identify such dogs before re-homing. In addition, all shelters want to maximise the likelihood of success of adoption and so attempt to identify those dogs most likely to succeed in their new home (van der Borg *et al.*, 1991) and not have behavioural problems such that they are returned to the shelter. A set of tests developed by van der Borg *et al.* (1991) predicted the likelihood of behavioural problems occurring in the new home. The tests had better predictive value (predict absence of behavioural problem) and sensitivity (correctly predict occurrence of behavioural problem) than the opinion of staff (Netto *et al.*, 1990). The tests were good at predicting leash pulling, aggression towards adult humans and dogs, disobedience and separation anxiety, but poor at predicting car-related problems (van der Borg *et al.*, 1991). The major problem with these tests is the time it takes to conduct them, so it is important that the tests be abbreviated but not lose their efficacy.

Aggression was a major reason for dogs being returned to shelters (Wells & Hepper, 2000b). A series of tests developed by Netto and Planta (1997) was found to be useful in assessing the aggressive tendencies of dogs. These tests were developed for use in breeding programmes but could be used, or modified and then used, in shelters.

An attempt by Hennessy *et al.* (2001) to correlate the behaviour of dogs in an animal shelter to their behaviour after adoption was not as successful as that of van der Borg *et al.* (1991), but different behavioural tests were used. The former found that pups that were quite fearless and had low plasma cortisol levels were more likely to have behavioural problems after adoption.

Shelters attempt to determine if the potential owner is likely to be a good owner and would keep the animal. The age and expectations of owners were important factors in determining whether an adoption would be successful (Kidd *et al.*, 1992a). An important factor in whether an adoption succeeds or not, is the new owner. Reasons why adoption from an animal shelter was unsuccessful included the new owner's ignorance about the (1) behaviour to be expected from a dog, (2) methods of training a dog, (3) time required to take care of a dog, (4) expense involved in owning a dog, and (5) parents with unreasonable expectations that a dog would teach their children to be loving, emotionally sensitive and responsible beings (Kidd *et al.*, 1992b). A higher percentage of men than women, first-time owners than previous owners of pets, and parents than non-parents, rejected the pet they had adopted from a shelter (Kidd *et al.*, 1992a)

Few people (4/41) (9.8%) who had recently acquired a new dog and were clients of a veterinarian rejected the new dog, when compared with the 20% of people who rejected a pet adopted from a shelter (Kidd *et al.*, 1992b). Of the four dogs rejected in that study, one was vicious, one could not be controlled by the wife, and two did not adapt to pets already in the house. These dogs were retained for about 6 months whereas those from a shelter were returned within 2 months. Veterinary clients had lower expectations of the dog than adopters from a shelter.

Shelters have different policies with regard to who they are willing to give dogs to (Balcom & Arluke, 2001). Closed policies are those which look for reasons not to allow adoption while open policies are less critical of potential owners and accept that a person seeking a pet will obtain one from a shelter or another source (Balcom & Arluke, 2001). The degree of preparation and planning for the dog by the new owner did not appear to influence the likelihood of a dog being relinquished (Patronek *et al.*, 1996), suggesting it may not be worthwhile for the shelter to invest too much time ensuring that the situation is optimal for the dog.

Irvine (2003) compared the way institutions dealing with unwanted pets portrayed them and the need for a commitment for the animal's lifetime. Such organisations may strive to help people keep their animals when the owners actually want to have troublesome animals removed from their home. Efforts to gain public support for help in dealing with unwanted pets often ignore the reality of the problem being dealt with.

8.   HOW SUCCESSFUL IS RE-HOMING?

The success of re-homing dogs from shelters is important (Table 7) and it is important to understand the reasons for success and failure of re-homing. In the Netherlands, the proportion of dogs returned to SPCA shelters increased from 19% in 1983 to 50% in 1991 (van der Borg *et al.*, 1991). In an American study, 18% of first-time adopted dogs were returned to the shelter (Patronek *et al.*, 1995). Following-up an adopted dog can be difficult as it may not be possible to contact the people who adopted the animal, especially as many people who adopt animals are young and may be transient (Phipps, 2003). In a study of a group of dogs de-sexed early in life, Howe *et al.* (2001) managed to contact the owners of 269 (22%) of 1,215 dogs. In New Zealand, Phipps (2003) contacted 103 (2.9%) of 354 persons who had adopted dogs within the previous 24 months. In Northern Ireland, 556 (37%) of 1,547 people who had purchased a dog from a shelter in the previous four weeks, responded to a survey by Wells and Hepper (2000b). In contrast, Kidd *et al.* (1992) contacted 343 of 392 people 6 months after they had adopted a pet.

The majority (68%) of the 556 respondents to a postal survey by Wells and Hepper (2000b) reported that their adopted dog had shown undesirable behaviour within four weeks of adoption. Common behaviours were fearfulness and overactivity, shown by 53% and 37% of the respondents' dogs, respectively. Male dogs were more likely to show aggression towards other dogs, have sexual problems and to stray than females, but the latter were more likely to be fearful. Pups showed fewer behavioural problems than juvenile or adult dogs. Dogs with behavioural problems were more likely to be returned to the shelter especially if they were aggressive towards humans, or exhibited separation anxiety (van der Borg *et al.*, 1991).

Table 7. The fate of re-homing dogs.

	USA[1]	USA[2]	USA[3]	N Ireland[4]	New Zealand[5]	Australia[6]
Number	269	1,229		556	100	4,405
Dog not still in home				12%		
Returned to shelter	13%	13%	12%	6%	10%	7%
Given away	6%					
Died				3%		
Ran away				>1%		

[1] Howe *et al.* (2001)
[2] Posage *et al.* (1998)
[3] Alexander and Shane (1994)
[4] Wells and Hepper (2000b)
[5] Phipps (2003)
[6] Marston *et al.* (2004)

Data available on what number of, and why, dogs are returned to shelters suggest ways and means of reducing the problem. Whether the problem with ownership is greater for dogs obtained from a shelter than from other sources is unclear, and it is not possible to compare these data with dogs bought or obtained from either pet shops or breeders as the data for the latter two sources of dogs are not available.

Factors which determine success of re-homing include matching the correct dog with the correct owner and then supporting the owner through the first few months of ownership. The latter is seldom possible, due to the staffing situation at many shelters. However, owners obtaining advice from a breeder, groomer, trainer, friend or neighbour was associated with an increased risk of relinquishment (Patronek *et al.*, 1996) in contrast to those seeking advice from a veterinarian. This may be related to the financial situation of the new owner. In one shelter, all potential owners were obliged to visit a private veterinarian within 10 days and compliance was greater than 96% (Patronek *et al.*, 1995).

Adopting out animals without ensuring the new owner has the financial ability to care for the dog may be a mistake. Thus, establishing how much they are able to spend on a dog may be sound information required for deciding whether adoption is worthwhile. De-sexing and training dogs that are likely to be re-homed may increase the success of adoption, as may having the new owner revisit the shelter after one week with their new dogs, for advice and help. The new home must also be appropriate in that it should not allow dogs to escape, especially those caught as strays.

Thus, it is important to identify dogs that are likely to be adopted and to ensure they are healthy and unlikely to engage in behaviours that will make them impossible to keep. It is also important that new owners know how much effort is required to keep a dog, a dog's normal behaviour, and how to train a dog, and that they do not to have unrealistic expectations.

9.   EARLY NEUTERING

The problem of unwanted pups and the difficulties of getting new owners of adopted dogs to neuter their animals encouraged re-homing agencies to start neutering animals before they were re-homed. This resulted in pups being castrated or spayed often at 8–16 weeks of age, a practice which started in the 1980s. In 1993, the American Veterinary Medical Association approved the practice in an effort to stem the overpopulation problem of dogs and cats. The practice has stimulated controversy and there is ongoing debate as to whether early neutering is acceptable with regard to the future health of the animals concerned.

In the USA, 70% of re-homed dogs and cats were not de-sexed despite contracts and encouragement from adoption agencies (Lieberman, 1988), and one state has made it mandatory that all animals from adopting agencies be neutered (Crenshaw & Carter, 1995). In Texas, USA, and Melbourne, Australia, fewer than 30% and 23%, respectively, of dogs in shelters were de-sexed (Mahlow, 1999; Marston et al., 2004). In one shelter, 53% of new owners did not get their dog neutered, and 43% of female and only 33% of male dogs were de-sexed (Alexander & Shane, 1994). This may be an international problem with only a small percentage of dogs being de-sexed after being adopted from a shelter. Only 10/248 (4%) of respondents in a survey by Wells and Hepper (2000a), who had adopted dogs from a shelter in Northern Ireland, had their male dog castrated, and 86.7% of respondents with a bitch did not know if it had been spayed. However, this may not be representative of the total population. In one survey, the majority (63%) of 1,335 dogs were neutered, and only 3.4% of 968 households reported that their bitch had a litter in the previous year, 66% of which were planned (Patronek et al., 1997). De-sexing appeared to increase the likelihood of an animal being adopted, and 73% and 71% of de-sexed dogs were adopted at two shelters in comparison to 36% and 45% of intact animals, respectively (Clevenger & Kass, 2003)

The castration of young male pups does not appear to have any long-term negative health effects. However, early neutering of females may lead to an increased risk of urinary incontinence and juvenile vaginitis, or result in an infantile vulva. There are two types of urinary incontinence, congenital and acquired. The congenital form resolves spontaneously in around 50% of bitches after their first heat although some bitches develop incontinence after their first heat. About 10% of neutered bitches developed urinary incontinence compared with 2% of intact bitches. Bitches neutered before their first heat were more likely to develop incontinence than those neutered after it (Thrusfield *et al.*, 1998). Professor Peter Holt, in a letter to the *Veterinary Record* (Holt, 2000), cautioned against the early neutering of a bitch pup that had signs of juvenile vaginitis, an infantile vulva, urinary incontinence, or was of a breed predisposed to urethral sphincter mechanism incompetence. These breeds include Old English Sheepdogs and Doberman Pinchers (Holt & Thrusfield, 1993). Holt (2000) observed that juvenile vaginitis and infantile vulvas resolved after the first heat. Caution such as that ascribed to Holt with regard to early neutering has been regularly supported (Jagoe & Serpell, 1988; Swift, 2000), but the case for early neutering of dogs has also been strongly supported by other veterinarians (Lieberman, 1988; Theran, 1993; Land, 2000; Anonymous, 2001c).

In a series of short-term studies, there were no differences in skeletal, physical or behavioural development between male and female pups neutered at seven weeks or seven months of age, 15 months after neutering (Salmeri *et al.*, 1991). Those findings have been supported by others (Crenshaw & Carter, 1995; Stubbs & Bloomberg, 1995). However, there was a delay in closure of the growth plate in pups neutered at seven weeks of age (Salmeri *et al.*, 1991), but the significance of this remains unclear (DeVile, 1998). In one long-term study, there were no differences in behavioural problems or in problems associated with any body system four years after either pre- or post-pubertal neutering of male and female dogs (Howe *et al.*, 2001). However, some are of the opinion that dogs that have been neutered before puberty have persistent juvenile behaviours than those neutered later (Lieberman, 1988). This has been supported by Sivacolundhu (1997), who found that pups neutered before five months of age were more likely to destroy things, bark or howl when separated from their owner, had shorter attention spans, and exhibited anxiety-based urination and defaecation. The possibility that early neutering has long-term effects on behaviour has been questioned by Hart (1987).

Surgical difficulties initially felt to be associated with early neutering have been overcome and many shelters now routinely have pups neutered at eight weeks of age, before re-homing. This practice has been recommended as a major step forward in reducing the production of unwanted pups, but data are not available to support this theory.

Being sexually intact was a major reason for surrendering dogs to a shelter in the first place but McCormick (1999) argued that de-sexing is a mutilation like tail docking or ear cropping. The counter arguments are that de-sexing prevents the production of unwanted pups and reduces the likelihood of pyometra, testicular and mammary cancers, and unwanted behaviours. In male dogs, castration reduced roaming, fighting or mounting (Hopkins *et al.*, 1976; Maarschalkerweerd *et al.*, 1997; Neilson *et al.*, 1997), but the presence of testosterone in intact dogs appeared to slow down cognitive impairment as the dog aged (Hart, 2001). Moreover, bitches may become more aggressive after spaying and develop an indiscriminate appetite (OFarrell & Peachey, 1990). In some countries such as Sweden, a small percentage of dogs are de-sexed but there appears to be no major problem with unwanted pups. Unwanted pups are produced by dogs belonging to a small proportion of dog owners, and these people should be targeted in dog control programmes. There are social and personal reasons for not having one's dog de-sexed, but responsible owners will not allow their dogs to produce unwanted pups. Owners who resist de-sexing their dogs may argue that intact dogs may have a much more interesting social existence and that this outweighs the possible production of unwanted pups. Responsible ownership does not equate with de-sexing but with controlled breeding.

10.   CONCLUSIONS

In many countries, the problem of unwanted dogs and pups is declining despite the increase in numbers of dogs. The media, movies (*Beethoven*, *K-9*, *Turner and Hooch*) and television shows (*The Simpsons*) support the non-relinquishment of dogs to shelters and pounds (Rajecki *et al.*, 2000), and concepts of responsible ownership are now probably widespread. This may be due to the work of animal welfare organisations, but increasingly restrictive legislation may be the driving force behind reduced numbers of stray and unwanted dogs. Dog-related legislation continues to become more restrictive and this is likely to further reduce the problem of stray and unwanted dogs and allow local authorities to focus more on the small percentage of irresponsible owners and deal with them accordingly.

The management of dogs in shelters and pounds is probably improving as our knowledge of dogs' requirements grows and attention focusses on how to maximise the success of adoption. Nevertheless, many unwanted dogs still enter shelters and pounds and are killed. The establishment and development of shelters in poorer countries is one of the most important ways that the unwanted dogs of the world can be managed, with attention paid to their welfare and the re-homing of suitable animals. However, this would require a substantial input from dog welfare organisations in wealthy countries and their supporters. Meanwhile, those working in canine welfare in poorer countries continue to work often under appalling conditions to care for unwanted dogs and to kill those that cannot be re-homed.

Chapter 11

BEHAVIOURAL PROBLEMS

Abstract: There are difficulties in defining the normal behaviour of dogs because of the
              different breeds, the difference between the behaviour of domestic dogs and
              the wolf, and the variance between the environments in which feral dogs and
              owned dogs live. Many dogs display behaviours which are normal but
              unacceptable to humans and restricting these behaviours limits the dog's
              ability to behave normally and express its normal repertoire of behaviours.
              Restriction may lead to abnormal behaviours that may be anxiety-based. There
              are genetic and ontogenic elements in the aetiology of abnormal behaviours.
              Abnormal behaviours may indicate that the management or environment of the
              individual dog, which expresses them, are inadequate. Some abnormal
              behaviours, such as continuous circling, may impact on the physical well-
              being of the affected dog. Providing an adequate enriched environment for a
              dog to stop engaging in these abnormal behaviours may be impossible. Thus,
              dogs may be put on lifelong medication to help them cope with the damage
              caused by an inadequate environment.

1.   INTRODUCTION

It is regularly stated that behavioural problems are the major reason for
dogs being killed in the USA (Overall, 1997). The veracity of this statement
depends on how the word 'problem' is defined. The normal behaviour of
juvenile dogs can make them less attractive than an 8–12 week-old pup, and
this may lead to young dogs being killed by the owner, surrendered to a
shelter, or to euthanasia by a private veterinarian. Many young dogs are
killed because of their behaviour but if a dog attains maturity its chances of
surviving to an old age are generally good. Many of the normal behaviours
of the dog are very attractive to people and explain why dogs are such

popular companions. Its social nature, playfulness and apparent love of family are attractive attributes. However, other normal behaviours make it less desirable as a companion and neighbour. Dogs are often very active and destructive; they scavenge, eat faeces, may defaecate freely and anywhere, they bark and bite, are sexually promiscuous and are predatory. The owner has to clean up after them on the street and control them in public. These unacceptable, though normal, behaviours and the owners' social responsibilities may lead to a dog being killed.

Some behaviours are abnormal and individual dogs who exhibit these may become dangerous to humans or self-destructive. These behaviours may be related to underlying anxiety, the aetiology of which may be complex, combining genetic, ontogenic and environmental factors. Treatment of these behaviours may be complicated and difficult for some owners and the dog may be killed or placed on lifelong medication. Behaviours are categorised as normal and acceptable, normal but unacceptable, or abnormal (Table 1). The acceptability of a behaviour may be defined by the owners, their neighbours, or society through public disapproval or legislation.

Table 1. The acceptability or not of different categories of behaviour

Behaviour	Category	Acceptable[a]	Unacceptable
Barking	Normal	Owner	Neighbour, society
Promiscuity	Normal		Owner, neighbour, society
Scavenging	Normal	Owner	Society
Defaecation in public	Normal	Owner	Owner, society
Territorial aggression	Normal	Owner, neighbour	Owner, neighbour, society
Dominance aggression	Normal		Owner
Tail chasing	Abnormal		Owner

[a] May vary according to specific conditions

The behaviour of a dog is important for its welfare because normal but unacceptable behaviour may result in the dog being abused, ignored, given away or killed. Abnormal behaviours presumed to be caused by anxiety, such as separation anxiety, are welfare concerns and result in a poor quality of life for the affected animals. There are a number of excellent textbooks on the aetiology, diagnosis and treatment of behavioural problems in dogs (Overall, 1997; Landsberg *et al.*, 1997; Lindsay, 2001). The impact of behaviour on the welfare of dogs will be discussed in this chapter.

2.   NORMAL BEHAVIOUR

A fundamental tenet of animal welfare advocates that an animal should have the freedom to engage in normal behaviour but it is difficult to definite normal behaviour. If defined as the behaviour of the progenitor species, then few if any animals under human management engage in normal behaviour. Indeed the owned dog is not allowed, by law and by animal welfare proponents, to engage in many normal-behaviours (Table 1). The dog has been defined as an abnormal wolf, behaving in a perpetually juvenile manner.

The normal behaviour of the dog might best be defined as that of wild or feral dogs such as the dingo. These animals live in small groups, pairs or as individuals as appropriate to their environment; they dig dens, hunt and scavenge, fight over territory, breed annually, defaecate freely and bark or howl as necessary. Free-living dogs in cities and rural areas worldwide behave similarly although they may breed more frequently. Many of these behaviours are permitted in rural and urban communities in the developing world but virtually none are acceptable in post-industrial wealthy, urban societies. Thus, by this definition, dogs are not allowed to behave normally in those societies where their welfare is considered the most. Indeed, many of the basic social communication techniques of dogs, such as smelling the perineal region or licking the lips of humans, are also unacceptable.

In addition to basic canine behaviours, there are breed-specific behaviours. Many breeds of dogs were selected over centuries to engage in particular activities and have a predisposition to these. Terriers are predisposed to hunt and kill small prey, and for those terriers still used and bred for this, such as the Jack Russell, hunting and killing is normal behaviour. Border Collies are bred to work sheep and will gather anything that moves, including humans. Both these breeds frequently engage in unacceptable behaviours, such as predating on cats and snapping at ankles, respectively. In many breeds, there are working strains and dogs from these are predisposed to engage in whatever work they were bred to do. When they cannot engage in such work, these animals are deprived of opportunities to engage in activities that are important to them.

An alternative definition of normal behaviour may simply be not engaging in abnormal behaviour. This is an inadequate definition, as many animals in what are agreed to be inadequate conditions do not engage in abnormal behaviours. Many sows in dry stalls and layer hens in battery cages do not engage in abnormal behaviours. Many dogs live in quite restrictive conditions but do not exhibit abnormal behaviours even when they cannot engage in some important normal behaviours.

A more acceptable definition of normal behaviour may be behaviour considered appropriate to the environment and acceptable to society. This allows humans to keep animals under a range of conditions appropriate to their own living conditions, and to use animals in ways acceptable to their society. This anthropocentric definition is also inadequate but probably the one used inadvertently by legislators there being no no simple alternative.

The majority of owned dogs living in houses and apartments learn to behave in an acceptable manner. But this does not mean that they are living adequate lives, only that they are living acceptable lives to their owners, their neighbours and the human society that allows them to exist. It may be impossible in such restrictive conditions for a dog to behave in a reasonably normal manner. This may be one reason why behavioural problems became more pronounced in the 1960s and 1970s when dogs moved off the street, often indoors into relative isolation, within smaller family units and with less time being spent with them.

Human expectations of dogs may be greater than dogs are able to provide. The emphasis in the media and scientific literature for the last three decades of how beneficial dogs are for children and for one's health and social life, and the desire of canine industries to promote pet ownership has encouraged dog ownership. The early work of Friedmann et al. (1980) was inadequately interpreted and overemphasised the health benefits of pet ownership. This may have led to many dogs being purchased as health enhancers. In the developed world, numbers of dogs increased in the 1980s and 1990s, perhaps in some degree as a response to the emphasis on dogs being a useful assistant with parenting and as a social lubricant. However, it was not often recognised how limited our ability to provide dogs with their behavioural requirements is. Dogs were obtained but not allowed to be dogs. Many dogs were surrendered to shelters when people refused to take responsibility for their animals or when they accepted that they could not provide what the animal required.

As the dog is a juvenile wolf it may need more social interaction than an adult wolf. Therefore, dogs may need more than one human or another dog for company. Indeed they may need to be in a pack situation, as wolf puppies and juveniles are in a litter, and receive a great deal of social contact throughout life. Many working dogs on large farms, Foxhounds, Beagles, and gundogs on large estates, live together in packs. Dogs living in groups in shelters may be more successful as pets after re-homing than those housed alone.

If dogs are not allowed to engage in normal behaviours and live in complex social environments then their welfare may be compromised. The importance of activities such as agility and obedience competitions may be that they allow dogs to be in large packs with complex social interactions, rather than the agility exercises *per se*. If the social and physical lives of dogs are inadequate then the question to be asked is should we allow dogs to live in these conditions at all or must we modify the dog's behavioural needs to those which we are able to provide with limited facilities. There may already be a public response to this question in those wealthy societies where the numbers of dogs is declining. People may have come to accept that they cannot provide what dogs need within their present lifestyles.

If the environment does not allow dogs to behave normally then some will engage in abnormal behaviour. These animals may have a genetic predisposition towards an inability to cope, and their experience as a young animal may be an important factor in the aetiology of the abnormal behaviour. The behavioural needs of the dog need to be defined but at present there is little evidence to show what these needs are. There have been no trials using demand curves trials to compare how dogs value essentials like different foods, social environments of different complexities, or different activities.

The pioneering work of Scott and Fuller (1965) defined four stages in the development of the young pup. The critical period, called the socialisation period, from 3 to 14 weeks of age, was a time when pups could cope with a wide range of novel environmental experiences and accept them in future as being of little significance. In the 1970s and 1980s, many owners identified that their dogs engaged in inappropriate behaviour and it was regularly suggested that this was due to the individual dog being isolated during the socialisation period, as recommended by veterinarians concerned about canine parvovirus disease. Owners complained about having disobedient dogs and dogs that were difficult to handle for grooming or medicating (Seksel *et al.*, 1999). To counter this, puppy socialisation classes were established in many countries to increase the social experience of pups in a safe environment and to educate owners about the care and training of their dogs.

Socialisation classes had little effect on the social responses of pups six months after the classes were attended (Seksel *et al.*, 1999). Ward (2003) quantified the experiences of pups in New Zealand and found that they were exposed to many different people and environments during the weeks after purchase and within the socialisation period. This might explain the lack of effect of socialisation classes described by Seksel (1999). If this is the case, then poor socialisation may not be a significant factor in the aetiology of many behavioural problems and other factors must be considered, including breed-specific behaviour, and the environment, including the owners' dog management ability, their behaviour and lifestyle.

Individual dog owners may misinterpret or ignore what their dogs are communicating, and may not know the motivation for a particular behaviour.

This may influence how the dog behaves and how it is treated. A lack of understanding of simple communication signs can lead to the inadvertent reinforcement of undesirable behaviours and the punishment of desirable activities.

3.   BEHAVIOURAL PROBLEMS

Many problem behaviours are normal behaviours but some are abnormal or pathological, either due to quantitative or qualitative differences from normal behaviour. Behavioural problems of dogs are obviously a serious concern to dog owners and veterinarians. In the last three decades, a plethora of books about dog training and behavioural problems of dogs have been published, so there is obviously a market for these texts and keen interest in the topic.

In the 1960s, many dog books related to sport and working dogs and their training. This changed in the 1970s, with more emphasis on training the pet dog and treating behavioural problems. The initial significant interest by the veterinary profession in the behavioural problems of dogs can be dated to the publication of the excellent text by Hart and Hart (1985b), titled "*Canine and Feline Behavioural Therapy*". This has been followed by several major texts in the last two decades, which discuss the prevention, diagnosis and treatment of behavioural problems of dogs. This may reflect a growing problem, but behavioural problems in companion dogs certainly existed

Table 2. Reports of behavioural problems with dogs.

Behaviour	No. dogs	% Complaints
Aggression	1,069	19
Excessive barking/vocalisation	741	13
Destructive chewing	628	11
Digging	501	9
Begging for food	471	8
House soiling	428	7
Jumping on furniture	361	6
Jumping on people	350	6
Running away	333	6
Fear especially of loud noises	301	5
Disobedience	171	3
Hyperactivity/excitement	171	3
Timidity	90	2
Stealing food or items	55	1
Destructive	34	1
Unacceptable eating habits	31	1
Eating faeces	13	0.2

Adapted from Beaver (1994)

before these texts were written. However, there may have been less interest in them by professionals and treatments were probably crude. Dog trainers working with working animals may use techniques that are not suited to companion animals.

In surveys, the majority of dog owners reported that their dog engaged in some unacceptable behaviours. In suburban Melbourne, Australia, 65% of owners stated that their dog had a behavioural problem (Kobelt, 2004), and in the USA 87% of dog-owning veterinary clients claimed the same (Campbell, 1986b). Beaver (1994) reviewed the literature on owners' complaints about canine behaviour, from nine reports totalling 4,487 people surveyed in Australia, North America and the UK. Results indicated that the most common problem was aggression, followed by barking, chewing and digging (Table 2). The common forms of aggression were identified as territorial and owner protection. When clinical cases were reviewed from the literature and from Beaver's own clinical records, the major problem remained aggression (Table 3), but the aggression was mainly dominance and fear-biting aggression. In addition, the other behavioural problems were quite different from those identified by surveying owners (Table 2).

Table 3. Major types of behavioural problems diagnosed in behavioural clinics.

Behaviour	No. diagnoses	% Cases
Aggression	1,083	63
House soiling	143	8
Destructive chewing	97	6
Fear/phobias	75	4
Hyperexcitability	48	3
Separation anxiety	34	2
Submissive behaviour	42	2
Excessive barking/vocalisation	32	2
Abnormal eating	20	1
Medical based problems	14	1
Others	130	8

Adapted from Beaver (1994)

Behavioural problems have been categorised by Overall (1997) into work-related, stress and anxiety, and those having a physiological basis. They are important in relation to the welfare of the dog because they may reflect an unpleasant state of existence, which we may name as anxiety or fear. The behaviour itself may reflect an abnormal mental state or it result in the animal being managed in a way that affects its welfare. Pathological behaviours, such as those due to excessive anxiety, and stereotypical behaviours reflect an abnormal mental state and dogs engaging in such behaviours are probably suffering some form of distress.

Behavioural problems are significant with regard to the welfare of dogs in that they may; impact on the physical and/or psychological well-being of the animal, be the product of an inadequate environment, reflect incorrect matching of the animal and environment, cause the human-dog relationship to deteriorate, increase the likelihood of the dog being abused or killed, affect the social mileau in which dogs live, and make it more difficult for the dog's owner and the owners of other dogs to exercise their dogs off the leash.

4.   FEAR AND ANXIETY

Fear is a primary emotion, which is useful for determining a response to a potentially dangerous stimulus. Fear may be acute or chronic but in all cases it is, by definition, unpleasant. A dog may be exposed to many potentially dangerous stimuli in its lifetime but it should learn to identify those that are dangerous and those that are not. Behaviour in response to fear is an adaptive response which allows for defensive reactions to promote survival. Fear becomes a significant welfare issue when the behavioural response is elicited by non-dangerous stimuli and it becomes excessively intense or overly long, or when the owner's response is inappropriate, for example, punishing submission urination. If fear occurs frequently, becomes chronic or leads to phobia or anxiety, then it is a welfare problem for the affected animal. This may affect the dog's health and alter or exaggerate its physiological responses to common environmental stimuli. An acute fearful response to a potential danger is not generally very significant as the response is usually brief and the dog can avoid the danger by moving away or freezing. Fear responses can decrease with gradual exposure to the stimulus by desensitization. Fear can be graded from normal to abnormal, with the degree proportional to the perception of the danger. A sudden profound all-or-nothing abnormal response that results in extreme behaviour is called a phobia (Overall, 1997). Phobias are excessive fears that are disproportionate to the danger of the stimulus that causes them (McCobb *et al.*, 2001).

Phobias develop quickly and do not decrease with repeated exposure to the stimulus. When a dog responds in an excessively fearful manner to a common and non-dangerous stimulus such as a man in uniform, then the animal's welfare may be compromised significantly. When this behavioural response prompts attention from its owner then it is reinforced and likely to increase, but if the behaviour is punished then the fear also increases (Landsberg *et al.*, 1997). Common stimuli that evoke extreme fearful responses or phobias in dogs include some types of people (men, men in uniforms, children), noise (gunfire, fireworks, thunderstorms) and particular places (veterinary clinics). Fearful responses to children or other people are common and easily treated by desensitization and counterconditioning.

The exaggerated fear response can result in direct injury to the dog as it frantically tries to escape from the stimulus through windows or doors, or rushes into the street. It can also have long-term physiological effects, though there is little research into this in dogs. In other species, chronic fear caused by poor human-animal interactions affects health and productivity (Hemsworth & Coleman, 1998) and there is no reason to doubt that chronic fear caused by poor owner-dog relationships cause health problems in dogs.

Noise phobias and panic attacks to loud or sudden noises are reasonably common, but such attacks have even occurred in response to birdsong (Aronson, 1999). These can be treated by medication and desensitisation and/or counter-conditioning. As some of these stimuli are seasonal, preventative pre-medication may be used to prevent panic attacks, but treatment to eliminate the response is better. In one survey, 40% of dogs suffering from thunderstorm phobia were under one year of age when they first showed signs, and dogs from animal shelters or rescue organisations were over-represented (McCobb et al., 2001). Dogs with thunderstorm phobia showed physical signs such as panting, shaking, salivation, loss of bladder and bowel control, and dilated pupils, and many sought human companionship, became destructive, vocalised, self-mutilated or hid (McCobb et al., 2001; Crowell-Davis et al., 2003). These dogs experienced extreme fear and often incurred injuries. Treatment is not always successful, although the use of a combination of drugs (clomipramine and alprazolam) plus behavioural modification resulted in an improvement in 30/32 dogs in one study by Crowell-Davis et al. (2003). A dog appeasement pheromone reduced the behavioural response of dogs to fireworks (Mills et al., 2003; Sheppard & Mills, 2003) and may be useful in the treatment of other phobias.

Anxiety is defined as the apprehensive anticipation of future danger (Overall, 1997), with increased vigilance, tension and often hyperactivity. Separation anxiety is defined as distress associated with separation from a preferred companion. It is diagnosed in 20–40% of dogs referred to animal behavioural practices in North America (Voith and Borchelt, 1996; Simpson, 2000), and it has been estimated that at least 14% of dogs presented at a veterinary clinic in the USA show signs of it (Overall et al., 2001). Dogs with noise or thunderstorm phobias are also likely to have separation anxiety (Overall et al., 2001). In dogs, the common signs of separation anxiety are destruction, excessive vocalisation, inappropriate elimination or self mutilation when the preferred person leaves the dog. This usually happens at home but can also happen in the car or elsewhere. Behaviours suggestive of separation anxiety are commonly identified by people surrendering dogs to animal shelters (Miller *et al.*, 1996). Dogs from animal shelters, found abandoned or adopted from veterinary hospitals tended to be more likely to suffer from separation anxiety than dogs obtained from breeders, friends or pet shops (Flannigan & Dodman, 2001). Dogs suffering from separation anxiety were very attached to their owner, followed their owner closely, engaged in excessive greeting and showed anxiety on a departure cue. Of importance, dogs from a home with a single adult (Flannigan & Dodman, 2001) were 2.5 times more likely to suffer from separation anxiety than dogs from homes with many people; sexually intact animals were three times less likely to be affected than de-sexed animals. Spoiling dogs, such as allowing them on the bed, or the presence of other dogs in the home, was not associated with separation anxiety (Voith *et al.*, 1992), but owners of dogs with the condition were more likely to punish verbally than physically and their dogs were less well trained and not willing to lie on command (Takeuchi *et al.*, 2001). Dogs that had been obedience-trained were less likely have separation anxiety than those not trained (Clark & Boyer, 1993). When fear and anxiety become significant behavioural problems in dogs there is probably a malfunctioning of the central nervous system. Psychotrophic drugs are commonly used in the initial treatment of these problems.

Dogs whose emotional, psychological and physical needs are met may be less likely to develop the overdependence underlying separation anxiety (Schwartz, 2003). These needs have not been defined and they probably differ between dogs. However, they most likely include social interaction, psychological stimulation and exercise. It may be that in a one-adult household it is difficult to meet the social requirements of individual dogs, particularly if the person spends much time away from the dog. Ppsychological stimulation in the non-working dog is probably based on complex social interactions rather than any other activity, and the fact that sexually intact dogs are much less likely to develop separation anxiety supports this premise. However, some dogs probably respond more intensely to particular stimuli. This hyper-reactivity may be pathological and an overt set of behaviours associated with a neurochemical dysfunction associated with anxiety (Overall *et al.*, 2001).

Dogs with separation anxiety are treated by establishing defined periods of dog-owner interaction, pre-departure exercise, enriching the dog's life physically and psychologically, and desensitising it to departure cues (Landsberg *et al.*, 1997; Overall, 1997), but drugs, especially tricyclic antidepressants, are often used as part of the therapy (Seksel & Lindemann, 2001). Drugs may have to be used for a prolonged period of time to overcome the problem. Prevention should target on enhancing the lives of susceptible dogs.

5. COMPULSIVE DISORDERS

In dogs, there is a group of behaviours, considered abnormal, which are usually classified as compulsive disorders, obsessive compulsive disorders, or stereotypic behaviours (Luescher *et al.*, 1991; Landsberg *et al.*, 1997; Overall, 1997; Luescher, 2003). These have been categorised into five types (Table 4), based on the nature of the presenting sign (Luescher, 2003). The cause of these behaviours is poorly defined but genotype, experience, medical problems, conditioning and stress caused by inadequate environmental conditions have all been considered important. Particular breeds appear prone to specific compulsive disorders. Flank-sucking is most frequently seen in Doberman Pinchers, circling in Bull Terriers, and tail-chasing in German Shepherd Dogs. Acral lick granulomas are most commonly seen after an injury to a leg. Dogs may engage in these behaviours because they have elicited attention from their owner. While genotype, experience and conditioning are important in the aetiology of compulsive disorders, it is thought by some that they are conflict behaviours caused by environmentally-induced conflict, frustration or stress (Luescher, 2003). Idiopathic aggression where the aggression is unprovoked and uncontrolled may be a compulsive disorder.

The development of compulsive disorders is probably a complex interaction encompassing many factors. In pigs and horses, the environment has been identified as a major factor in the development of stereotypic behaviours. Enhancing social contact and changing feed type reduced stereotypic behaviour in horses (Cooper & McGreevy, 2002), as did increasing feed volume in pigs (Terlouw & Lawrence, 1993). Specific environmental factors have not been identified as being important in the development of compulsive disorders in dogs, although Scottish Terriers, when isolated as pups, developed whirling behaviour (Thompson *et al.*, 1956). Luescher (2003) suggested that any environmental factor resulting in frustration, for example no exercise off the property, conflict behavioiur such as inconsistent interaction between owner and dog, or stress such as that caused by other stressful behaviours, could contribute to a compulsive behaviour. However, the physiological responses of dogs with compulsive behaviours is poorly understood. As these experiences listed above are common, it could be expected that compulsive disorders would also be common, and they are probably under-diagnosed. Problem behaviours such as persistent barking or digging might be compulsive disorders but may not be diagnosed as such.

Table 4. Categories of compulsive behaviours, based on presenting signs.

Behaviour	Signs
Locomotion	Circling, tail-chasing, chasing light reflections, jumping in place, pacing
Oral	Leg- or foot-chewing, self-licking, air- or nose-licking, flank-sucking, polydipsia, polyphagia, pica, snapping in the air, scratching/chewing/licking objects
Aggression	Self-directed aggression such as growling at or biting the rear end, hindlegs or tail; attacking food bowl or other inanimate object; possibly unpredictable aggression towards humans
Vocalisation	Rhythmic barking or whining, persistent howling
Hallucinatory	Staring at shadows, chasing light reflections, startling

Adapted from Luescher (2003)

Compulsive disorders are almost certainly indicative of poor welfare, both in their aetiology and in the fact that they may interfere with normal behaviour. They may be self-destructive. Treatment is difficult as it combines environmental and behavioural modification with drugs. Environmental modification involves making owner-dog interactions more consistent and giving the dog more exercise and work. Environmental changes alone do not usually overcome the problem and the long-term administration of tricyclic antidepressants, such as clomipramine (Hewson and Luescher, 1998; Seksel & Lindeman, 2001), is required (Overall, 1997). The prognosis for complete recovery is influenced by the duration of the problem, but lifelong drug therapy is not uncommon (Overall, 1997; Luescher, 2003).

The severity of compulsive disorders and its intractable nature suggests that something truly abnormal is happening in both the breeding and the management of some dogs. It is likely that the environmental inadequacy underlies the problem in dogs is social in origin. It may be that affected dogs have little to do, no work, little significant social interaction with other dogs, little exploration, and inadequate interactions with their owners. The development of compulsive disorders is indicative of many dogs leading poor, miserable lives, despite being cared for by their owners, and may suggest that the lives of many pet dogs are inadequate.

The degree to which the welfare of individual animals with a compulsive disorder is compromised is unknown and will presumably be influenced by the developmental stage of the disorder and the type of disorder. Some behaviours may be signs of a psychopathology (Duncan *et al.*, 1993). Individual dogs may be stressed while engaging in the behaviour but for others it may be a coping mechanism (Duncan *et al.*, 1993), to deal with an inadequate environment. Mason (1991) compared a stereotypy to a scar in that it tells us something about the past. The existence of a compulsive disorder suggests that the environment in which the dog lived was inadequate and that its welfare was probably inadequate at some time.

The proportion of dogs with compulsive disorders is unknown and difficult to estimate but it is probably under-diagnosed. The risk factors and background to individual compulsive disorders need to be clarified for us to understand what exactly is deficient in the environment.

6.   AGGRESSION

Aggression in dogs has been classified in several ways. A functional classification of aggression includes fear, dominance, territorial, play, protective, idiopathic, possessive, redirected, food-related, inter-dog, maternal and predatory (Overall, 1997). In all forms of aggression, the aggressive dog can be injured as can the victim, be it human, canine or other species. Aggression is significant for the welfare of dogs in that it may result in dogs being killed (Reisner *et al.*, 1994; Galac & Knol, 1997) rather than treated, and prevention of its development increases a dog's chance of survival. In one German state during 1990, 34 dogs were shot by police after the dogs endangered, injured or "lethally injured humans". The dogs included 16 German Shepherd Dogs, five Pit Bull Terriers, three Boxers, two Rottweilers and two Bernese Mountain Dogs (Roll & Unshelm, 1997). In addition, the public belief, or the reality, that dog attacks have become a significant public health issue influences national and local politicians whose usual response is to make dog ownership more difficult and dog control in public more strenuous. Legislation may state that dogs have to be on a leash at all times in public places and never allowed to run free, or that certain breeds be muzzled at all times in public, be neutered and even killed. In the UK, the Dangerous Dog Act of 1991 did not seem to reduce the number of dog bites to humans over the following 2 years, and even tighter dog control legislation was proposed (Klassen *et al.*, 1996). In the Netherlands, behavioural testing for aggression was suggested as one way breeders might reduce aggressive tendencies in their dogs (Netto & Planta, 1997).

In many forms of aggression the dog protects itself (fear), its territory (territorial), or something or someone of value (protective, possessive), food, or pups (maternal). The fear involved in self-protection may not be significant in that it may be temporary rather than chronic. However, some forms of dominance aggression may be caused by ongoing anxiety about the dogs own position in the hierarchy and constant attempts to define its status (Overall, 1997). This type of aggression is likely to have long-term effects on the dog's well-being, in addition to the danger of the dog being dispensed with by owners unable to cope with the problem. Most emphasis on canine aggression has focussed on aggression towards humans (Langley, 1992; Poberscek & Blackshaw, 1993; Sacks & Lockwood, 1996; Stafford, 1996). In the UK, the incidence of dominance aggression may be decreasing, as the Association of Pet Behaviour Counsellors report for 2000 (Anonymous, 2001b) suggested it had decreased from 10% to 5% of their consultancy from 1996 to 2000. However, fear-related aggression towards other dogs had increased from 6% in 1996 to 14% in 2000. This type of aggression does not make the television news, but is significant from a dog welfare perspective, as both victim and aggressor may suffer. In an Australian survey, 36% and 48% of 105 respondents admitted that their dogs had bitten people or fought with dogs, respectively (Adams & Clark, 1989), and aggression towards other dogs was common in dogs purchased at a shelter (Wells & Hepper, 2000b). Dog-to-dog aggression can be categorised as fear, territorial, dominance, protective, possessive, food-related, maternal, play, redirected or even predatorial. It may occur within the home or in public (Landsberg *et al.*, 1997; Overall, 1997).

In one German study, dogs that were victims of aggression in public places by other dogs were usually walked by women (Roll & Unshelm, 1997). Most of the dog fights occurred when both dogs, aggressor and victim, were off the leash, but 14% of aggressors and 35% of victims were on the leash when a fight occurred. The aggressive dogs were usually owned by men, typically 30–39 years of age, who often had the dog for security reasons. Canine victims of dog attacks may become fearful and aggressive towards all dogs or dogs of a particular type. Much of the discussion about Pit Bull Terriers has ignored their predisposition to attack other dogs (Bandow, 1996), but this makes them unsuited to most environments where people and dogs live together. Most forms of dog-to-dog aggression are normal, but unacceptable. These forms of aggression by themselves may not be a particularly significant factor in the welfare of either dog involved, as it is normal behaviour and injuries are usually moderate rather than severe, but the response of humans to such aggression may cause significant restriction in the lifestyle of either the aggressor or victim, and this may have a more significant impact on the dog's welfare than the aggression itself.

7.   CONCLUSIONS

The normal behaviour of the dog may be unacceptable to its owner and/or society and may result in the dog being killed. However, such normal behaviour is unlikely to be damaging to the dog directly, and the welfare significance is the human response to it. An inability to engage in normal behaviour may result in the development of abnormal behaviours and these may be caused by chronic anxiety. Anxiety-based disorders have a physiological mechanism, but many are caused or exaggerated by different aspects of the dog's management. The apparently increased prevalence of behavioural problems in dogs in the late 20th Century may or may not be a real increase. If it was, it was probably amplified by changes in the dog's lifestyle in order to accommodate societal requirements and those of owners. The new lifestyle of dogs may be inadequate for a social, exploratory animal now confined indoors, isolated from other dogs and often given limited exercise either on a leash or free-running. Many aspects of modern urban living do not allow for the behavioural needs of the dog.

Chapter 12

THE PET DOG

Abstract: In Europe and North America, the majority of dogs are kept as companion
              animals. It is assumed that dogs owned as pets have a better welfare than
              working dogs or dogs used in research. However, the welfare of these pet
              animals will vary depending on the attitude, dog handling ability, economic
              status, and lifestyle of their owners, and the physical environment they inhabit.
              People might provide physically for their companion animals but may not
              provide them with an environment that is sufficiently complex mentally and
              socially. Many dogs live alone for the greater part of their lives and this may
              cause anxiety. The close relationship between humans and companion animals
              can result in greater expectations on behalf of the owner than the dog can
              provide, and disappointment may result in the dog being ignored, abandoned,
              or abused. Dogs are often given as presents to people who do not want them.
              Dogs may be fashion statements or used for support by people with inadequate
              social lives or paranoia. Dogs living in back yards may be isolated and
              ignored. However many pet dogs live good lives with their human
              companions.

1.   INTRODUCTION

The majority of dogs living in Europe, North America and Oceania as
pets are now often classified as companion animals. This category of animal
includes cats, small mammals (rats, mice, hamsters, guinea pigs), sometimes
horses and some birds. However, the dog and cat are the leading companion
animals and this status has led to major improvements in the welfare of the
dog as it has become a close companion and family member. The ratio of
dogs to humans varies widely in the developed world (Table 1); the USA,
Australia and France have the most dogs, and Germany and Japan the least.

The favoured dogs in Japan are small breeds while in the UK 43% of dogs are large or giant-sized breeds. The major improvements in canine welfare are principally physical in that their nutrition, physical environment and health have all improved dramatically in the last few decades. However, the question remains as to whether their psychological well-being has also improved.

Table 1. Human populations and the ratio of dogs to humans (adapted from Marsh, 1994).

Country	Human population (millions)	Dogs/100 people
Australia	17.4	18
Belgium/Luxembourg	10.4	13
Canada	25.9	12
Denmark	5.1	13
Europe (average)		11
France	56.3	18
Germany	79.3	6
Ireland	3.5	14
Italy	57.9	9
Japan	122.1	6
Netherlands	14.9	9
Portugal	10.4	13
Spain	39.0	10
United Kingdom	57.4	13
United States of America	252.1	21

It is often assumed that dogs owned as companion animals have better welfare than those used as working animals, athletes or in research. This may not be the case, as the value, management and treatment of a pet dog will vary depending on the attitude of owners and their ability to provide for the dog. There have been few efforts to assess the welfare of dogs living as companions with humans, and it may be difficult to do so because of the wide variation of circumstances under which dogs share homes with humans (Hubrecht, 1995a). The physical conditions under which pet dogs live probably vary more than that experienced by working, athletic or research dogs. Some pet dogs live indoors in an apartment or house virtually all the time, others in the back yard, others outdoors and indoors, while others are allowed to stray. Recently, Kobelt (2004) attempted to assess the physiological status of dogs living in back yards in Melbourne, and Beerda *et al.* (2000) compared the physiology and behaviour of owned dogs and laboratory dogs in the Netherlands. The results of these two studies suggest that there may be welfare problems with pet dogs living in two quite different environments.

The keeping of dogs as companion animals varies with the demography of the household. Families with school-aged children were more likely to have pets, including dogs, than families with infants, or older couples whose children have left home (Albert & Bulcroft, 1987, 1988). Large families were more likely to have dogs. In families with children, emotional attachment to the dog was lower than that felt by people living alone or couples without children. In childless families, dogs and people interacted more than in families with children. Thus, depending on circumstances the family dog may get a lot or very little attention and be of high or low value. Owners living in close physical contact with their dogs may become more and more attached to them.

Many surveys suggest that owners consider the dog to be a family member, but what does this really mean? In many of those surveys, this stance was expected as it is the socially correct answer; besides how else should a companion animal be categorised? Being a family member does not necessarily mean much and more effective determinants of attachments might be the amount of non-working time people spend with their dog, and the percentage of income and nursing time they would spend on the dog if it became ill. These determinants are rarely assessed, but in the USA and Australia only about 66% of respondents of a questionnaire took their dogs for a walk (Slater *et al.*, 1995; Kobelt, 2004), and in the USA few were willing to take out pet health insurance or pay more than $1,000 for veterinary fees.

The majority of pet dogs live in single-dog homes. In UK, 77% of households with dogs had one dog and the rest had two or more (Anonymous, 1998b), figures similar to those in suburban Melbourne (Kobelt *et al.*, 2003b). Social deprivation is considered a more significant deprivation for laboratory dogs than spatial limitation (Hetts *et al.*, 1992). If, as in many families both partners work, then the pet dog may live a socially-deprived life. Dogs in families with children were less likely to be over-excited, engage in inappropriate elimination, chew and escape than dogs in childless families (Kobelt, 2004). This may be due to owners with children being more effective at dog training and control, but it suggests that the dog has a more exciting social life with children coming home from school and spending quality fun time with them. Dogs may be beneficial for children, but children are probably even better for dogs and the larger the family the fewer behavioural problems dogs have (Campbell, 1986a).

Although rural families are more likely to own dogs, many dogs live in cities with limited access to areas where they can exercise off the lead. There are examples of societies with small populations of dogs living in large cities. Tokyo has about 12 million people, and dogs are regularly seen being walked on its streets. But, one rarely sees dog faeces, and dogs generally ignore passersby and appear well-behaved. These dogs may never run free off their owner's properties; they appear to always be on a leash. Is this acceptable or do dogs need to run free and engage in all those other outdoor activities people associate with the animal? In this chapter, aspects of the welfare of pet dogs will be discussed.

2.   THE PET DOG AND ITS OWNER

In evaluating his life, the prince in Guiseppe di Lampedusa's (1958) novel. The Leopard considered that the ownership of a number of dogs had been a significantly positive aspect of his life. "Yes, much on the credit side came from … the dogs; Fufi, the fat pug of his childhood, the impetuous poodle Tom, confidante and friend, Speedy's gentle eyes, Bendico's delicious nonsense, the caressing paws of Pop, the pointer at the moment searching for him under bushes and garden chairs and never to see him again". This is the experience of many people. However, another perspective on dogs is shown below in Beaver's (1997) analysis of dog ownership in the USA. Dogs are focussed on human behaviour (Hare & Tomasello, 1999; Naderi et al., 2001; Hare et al., 2002), are attached to humans (Prato-Previde et al., 2003), seemingly solicit help from their owner (Topal et al., 1998), and apparently have learned to communicate with humans by barking (Douglas, 2004) and via specific play mechanisms (Rooney & Bradshaw, 2002). This human focus may be common to other domestic animals such as cattle and pigs but has not yet been investigated in those species. However, such a focus suggests the possibility that dogs need and benefit from human contact (Odendaal & Meintjes, 2003) and interaction. Beaver's (1997) observations in the USA suggest that the human response to this need is inadequate. In the USA, each year, 1–3 million people are bitten by dogs; 1 in 200 people are bitten by 1 in 20 dogs; 500,000 people require medical attention because of dog bites; almost 50% of children are bitten before they are 18 years of age; 10–15 people die each year because of dog attacks; more than $1 billion is spent on dog-related liability claims each year; 20–25% of the population of dogs is surrendered each year to shelters because of behavioural problems; euthanasia is the leading cause of death among companion animals; and animal issues generate more input to legislators than almost any other social issue.

People usually report that they have a dog for companionship (Table 2), but for what other reason would one have an animal that does nothing all day? As a companion, dogs may help relieve stress and have a positive effect on the physical and psychological health of their owners (Hart, 1995). In some countries, the majority of people have dogs primarily for reasons other than for companionship. In the UK and USA, 58% and 70%, respectively, owned a dog primarily for companionship, but in Taiwan it was 41%, less than the percentage (47%) that had a dog for guard work (Hsu *et al.*, 2003).

Table 2. Reasons for owning a pet.

Reason	Importance			Total[a]
	Very important	**Important**	**Not important**	
Companionship	156	93	17	671
Love and affection	149	98	15	658
For the children	106	74	63	528
Someone to greet me	76	101	75	505
Property protection	60	96	95	467
Someone to care for	41	95	114	427
Beauty of animal	43	86	115	416
Sport, e.g. hunting	5	11	222	259
Show value	1	6	234	247

[a] Total score when weighted 1, 2 and 3 for not important, important and very important, respectively
Adapted from Leslie *et al.* (1994)

Many people have dogs for their children, and families with school age children were more likely to have dogs than families with preschoolers or without children (Leslie *et al.*, 1994). Kidd and Kidd (1989, 1990) reported that women were more attached to dogs than men, childless couples more attached than parents, and that 25% of parents expected their pet to teach their children responsibility. Kidd *et al.* (1992a) found that parents expected their pet to keep their children busy; dogs were expected to be companions, confidantes and sources of emotional support. People who did not have dogs cited housing difficulties as a major reason (Selby *et al.*, 1980), but the bother caused by having pets and wanting to travel were also important (Table 3). Families with no children or with small children may travel more than those with school-aged children which probably impacts on pet ownership (Leslie *et al.*, 1994). In one study of pre-adolescents, Davis (1987) found that families acquired a pet because of a 'pet deficit', that is the need for a pet to make their environment feel complete. This filling of a deficit may not bode well for the welfare of the dog as it is not being obtained for a definite reason, only to fill an apparent need.

Table 3. Reasons for not owning a pet.

Reason	Importance			Total[a]
	Very important	**Important**	**Not important**	
A problem when I go away	50	37	19	243
Not enough time	36	28	36	200
Poor housing for pet	39	16	49	198
Location dangerous for pets	26	25	47	175
Pets not allowed	28	7	68	166
Family allergy problem	19	15	61	148
Dislike animals	17	6	74	137
Too expensive	4	30	63	135
Zoonoses	9	21	62	131

[a] Total score when weighted 1, 2 and 3 for not important, important and very important, respectively
Adapted from Leslie *et al.* (1994).

In Ontario, families with preschoolers and urban dwellers were less likely to have pets than families with older children or rural dwellers, all of which may be good for the welfare of the dog. There, socio-economic status had no effect on pet ownership (Leslie *et al.*, 1994) but in New Zealand, lower household employment reduced the likelihood of dog ownership (Fifield & Forsyth, 1999). There is little evidence that higher incomes result in the family pet being better off with regard to its welfare and in the Netherlands the higher the income the more likely the dog was to have behavioural problems (Endenburg & Knol, 1994). Higher incomes should allow more money to be available for veterinary care and food, but more time may be available for the dog in families where there are unemployed people. Overall, dogs living in single adult, childless households tended to have the lowest incidence of behavioural problems, whereas problem-free dogs were more common in household with three or more children (Campbell, 1986a), which may reflect children keeping the dog happy and busy (Millot *et al.*, 1988).

The widely-published evidence that dogs are beneficial for a human's health and social life (Hart, 1995) may have led to an increase in the ownership of dogs during the 1980s and 1990s. Likewise the widely-heralded value of dogs for children (Paul & Serpell, 1996) may have encouraged parents to obtain dogs for their children without considering the requirements of the animal. In several studies, dogs sourced from shelters 'for the children' were frequently returned (Kidd *et al.*, 1992a; Phipps, 2003), suggesting that adopters had not thought through the consequences of adoption and the limited ability of dogs to assist them rear their children! In addition, children are regularly bitten by the family dog. However, preventing such bites by fostering relationships between children and dogs based on mutual deference, respect and communication (Love & Overall, 2001) will improve the welfare of the family dog by making it into a much-loved companion.

Dogs are now being recommended, indirectly, to fill a void in peoples' lives. The evidence that dogs are beneficial for the psychological, social and physical health of people could be used to promote the ownership of dogs by people living alone, anti-social or shy people, or for people who are afraid. People who have dogs and who benefit from their company may be different from those who choose not to have them. Widespread publication of the advantages of dogs for people may result in dogs being acquired by people who will not benefit from their companionship, resulting in dogs in homes where they are not really wanted. More worrying is the purchase of dogs to protect people. This may result in the dog becoming dominant and dangerous to owners who may have poor self esteem.

Although seldom identified in surveys, dogs of particular breeds may be acquired because they are fashionable. The popularity of the Siberian Husky in the 1990s is a good example, but this breed may be inappropriate as a pet. Dalmatians became popular after the movie *101 Dalmatians*, Border Collies after *Babe*, and apparently Akitas made the headlines during the OJ Simpsom trial in the USA (Inserro, 1997). A specific breed may go out of style quickly. Individual dogs of breeds that become unfashionable or are inappropriate may be lucky and get re-homed through breed recovery clubs or shelters, but many will be killed.

There is controversy as to whether human attitudes and behaviour influence the incidence of behavioural problems in dogs. O'Farrell (1997) found evidence of an association between dominance aggression in the dog and emotional anthropomorphic involvement with the dog, but Voith *et al.* (1992) did not find this. It is intuitive that in a close relationship between human and dog that the behaviour of one will influence that of the other. Dog ownership changes the behaviour of owners and it is likely that an owner's behaviour will influence a dog's behaviour. The question is whether increased intimacy in the human-dog relationship has negative effects on the welfare of the dog. If human attitudes increase dominance aggression, over-excitement, and unacceptable activities such as mounting people or inanimate objects, destruction if left alone, excessive attention seeking or biting people, as demonstrated by O'Farrell (1995, 1997), then anxious or overly anthropomorphic owners may have a negative effect on their dog's welfare. This also suggests that people who are not so anxious or overly anthropomorphic about their dogs do not tend to have dogs with behavioural problems. Thus, the people who may feel most concerned about their dog's welfare are not necessarily the most fit to have one.

The failure of the human-dog relationship has received much attention in the last few decades, especially with regards to the surrender of dogs to animal shelters and the return of dogs re-homed from such shelters. The owner's perception and tolerance are extremely important in whether dogs are kept or are dispensed with (Ledger, 1997). Ledger (1999) suggested that mismatching can be avoided if the dog's behavioural and physical characteristics are those considered desirable by the potential new owners. This supports the concept that the closer the dog's behaviour is to the owner's ideal the more attached the owner is to the dog (Serpell, 1996). The reason for obtaining a dog may be of great importance in whether a dog remains or is discarded. As shown in Table 2, there are some poignant reasons (someone/something to give me love and affection and to greet me) for acquiring a dog. For some people this may mean some inadequacy in their human relationships, and loneliness. Owning a dog to herd sheep and owning a dog to provide love and affection are quite different. Dogs will, of course, appear to provide affection, but is it correct to expect them to provide it and, indeed, little else? In Table 2, there is no indication that people own dogs in order to provide the dog with an active and exciting life; the reasons are all egoistic, that is what I want a dog to do for me. Although most owners state that companionship is the major reason for having a dog, in one study almost a third of owners had not played or spent time interacting with their dogs in the previous week (Case, 1987) so there appear to be marked differences between what people say when surveyed and what they actually do. This lack of attention is of concern, as human interaction with dogs can reduce stress and probably give pleasure in the latter (Hennessy et al., 1998; Odendaal & Meintjes, 2003). Of interest is the suggestion by Hennessy et al. (1998) that dogs in shelters are extremely sensitive to small differences in the quality of style of human interaction, based on the observation that these dogs had a lower plasma cortisol response when petted by women than men.

Wilbur (1976, cited by O'Farrell, 1997) interviewed 350 dog owners and by cluster analysis identified five classes of dog owner: 'companion' owners (27%), who considered the dog to be a family member; 'enthusiastic' owners (17%), who enjoyed their dogs but were not closely involved with them; 'worried' owners (24%), who were attached but often worried about their dog's behaviour; 'valued object' owners (19%), who regarded the dog as a possession and had no emotional attachment; and 'dissatisfied' owners (19%), who regarded the dog as a nuisance. If these figures are still pertinent and if human anxiety influences canine behaviour, then the welfare of a large percentage of dogs is of concern.

Ignorance and lack of experience of owners may also impact on the incidence of behavioural problems in dogs, especially with first-time owners (Jagoe & Serpell, 1996). However, it may be that first-time owners report problems because they have greater expectations of the dog and less knowledge of what to expect. Lack of tolerance, of normal dog behaviour and physical features (hair loss, smell) appears to be a significant factor in abandonment.

There is a need for more information on the detail of human-dog interactions in the home and whether dogs actually are afforded much attention at all. Play is considered an essential component of the dog-human relationship, and dogs are one of the few mammals that engage with other species in play. The motivation in dogs to play with humans is not reduced by playing with other dogs, and dog-dog play is different from dog-human play (Rooney *et al.*, 2000). This suggests that dog-human play is important for dogs regardless of the opportunity to play with other dogs. The outcome of tug games had no effect on the dominance element in dog-human relationships (Rooney & Bradshaw, 2002), but if play signals were absent or misinterpreted by dogs then potentially dominant dogs might use games to assert themselves (Rooney *et al.*, 2002).

3.   REARING PUPS

Dogs are obtained from many sources. In Ontario, 44% of dogs in rural communities came from friends and family, 40% of urban dogs came from breeders, while 8% and 11% of rural and urban dogs, respectively, came from shelters (Leslie *et al.*, 1994). Another source of dogs includes pet shops. People who chose a pet for themselves were more likely to bond closely with it than to a pet given as a present (Kogan & Viney, 1998). Pet shops and shelters are not ideal places to rear pups and prepare them for life in the average family. They pups may be confined, and be unable to go to the toilet except in the small cage, which later may make them more difficult to house-train. They may not be exposed to a wide variety of environments they will encounter later in life. Puppy farms produce pups for the pet market and the size of the market determines the likelihood of such farms developing. Puppy farms are not bad for the welfare of pups *per se*, but if there are too many pups and too few people to mind them then their socialisation and physical health will be compromised.

It is frequently stated that the experience of pups influences their future ability to cope (Serpell & Jagoe, 1995), and that inadequate exposure during puppyhood to a wide variety of environments may result in dogs that are fearful and anxious. Apparently, pups that did not experience particular stimuli during their socialisation period (from 4 to 14 weeks) were more likely to develop a fear response to them as adults (Appleby *et al.*, 2002). However, little is actually known about the relationship between the experiences during puppyhood and the development of behavioural problems during adulthood.

Handling neonatal pups, one to two weeks of age, for a few minutes each day and exposing them to mild environmental stressors, such as movement and temperature changes apparently improves learning ability and reduces emotional reactivity later in life. Landsberg *et al.* (1997), Overall (1997) and Lindsay (2000a) recommended exposing pups to a wide variety of environments. However, a study by Ward (2003) found that most pups in New Zealand were exposed to a large number of people, dogs and environments during this period without any special effort on behalf of their owners. Seksel (1999) found that the attendance at puppy socialisation classes did not lead to any change in the dog's later responses to novel, social or handling stimuli. This suggests that the majority of pups are exposed to sufficient variation in their environment so as to be able to cope in the future. However, pups adopted from a shelter were more likely to be retained in their adoptive home if they attended socialisation classes (Duxbury *et al.*, 2003). This suggests that pups from shelters do not have the broad environmental experience that those from other sources do, and suggests that shelter staff need to do more to 'socialise' their pups before adoption. It is recommended to breeders that they handle and expose their pups to a wide range of experiences before sale.

Pups that experienced a non-domestic maternal environment and a lack of urban experience between 3 and 6 months of age were more likely to be aggesssive towards unfamiliar people (Appleby *et al.*, 2002). This implies that the experience of pups during the first 6 months of life, before and after weaning, is important in developing confident and safe behaviour.

It is now generally recommended that pups be weaned at 7–8 weeks and, if possible, be sold or re-homed at this time. Weaning a pup earlier than 7 weeks may be detrimental to its development as it needs time to socialise with other dogs. Overall (1997) stated that 6 weeks was too early to separate a pup from its litter, due to profound behavioural responses at this time (Elliot & Scott, 1961), and because it caused recidivistic changes in the pup's behavioural development. Many puppies enter new homes at 7–8 weeks of age, but even if they remain with the breeder for longer periods this is not detrimental, if the pups experience a wide variety of environmental conditions, perhaps travel by car and meet a range of people, especially children.

Puppy socialisation classes have positive effects in educating owners about training their dogs, and initiating training of the pup (Seksel *et al.*, 1999), and may have some positive effects on adult dog behaviour (Ward, 2003) and retention of pups (Duxbury *et al.*, 2003). The latter may be due to better education of owners rather than any change in the pup's behaviour *per se*.

4.   INDOORS AND ISOLATED

Most companion dogs live as single dogs with one or more people. Many of them live indoors or outside in a back yard. Generally, pet dogs are inactive (Kobelt, 2004) and they are not doing what their breed was originally produced for. Many spend the majority of their lives alone except for some brief attention in the morning or evening from their owners. The owner may leave for work at 0700 or 0800 hours and return at 1800 hours and the dog, which is primarily a diurnal animal, is alone for the time when it is normally most active. This enforced isolation and inactivity must have negative effects on the welfare of some dogs. This aspect of the companion dog's life is poorly investigated, but the frequency with which owners complain of their dog behaving inappropriately suggests that there are many problems with dogs which may be associated with this isolation.

Dogs have been selected to focus on human behaviour. They probably need to interact with other dogs and humans. Dogs benefit from close interactions with their owners, and it results in an increase in plasma dopamine concentrations and other biochemical indicators of pleasure, and reduced stress (Odendaal & Meintjes, 2003). Dogs left alone for most of the day may be more likely to develop anxiety problems than those living in the company of either other animals or people. In laboratories, dogs penned with another dog were similar physiologically to those penned alone suggesting that dogs may need to live in larger groups and perhaps to range and explore like owned but free-roaming dogs.

The quality of life of many loved, single dogs bred to be active and working, but restricted and isolated is definitely inadequate, as is the life of many that are not loved but simply owned. The social pressures which have resulted in dogs being moved indoors from the streets, and the demise of the latch-key dog, may have been an improvement in our relationship with dogs but may not have improved the dog's life that was subsequently limited to being indoors or in a back yard. The pressure to own one dog and the limited space to own more may have had serious implications on the welfare of the family dog. It is certain that many owned dogs do not live as fortunate a life as that of research dogs described by Loveridge (1998). It is widely accepted that caging dogs alone results in under-stimulation and the development of behavioural problems. The diagnosis of behavioural problems in many owned dogs living inside or outdoors in a back yard suggests that they are living dull, uninteresting lives.

The common advice given to new dog owners is to keep their dog active, engage it in obedience or agility, walk it frequently, and interact with it in a manner that keeps the dog busy physically and mentally. These recommendations are not possible for many people and whether obedience, especially competitive obedience, or competitive agility, is actually beneficial for dogs is debatable. However, giving a dog a substantial walk had noticeably positive effects on dogs held as individuals in laboratory cages (Beerda et al., 2000), and this suggests that the experience of being brought for a walk is a major method of reducing stress in laboratory dogs, and probably pet dogs also. In that study, owned dogs housed during working hours in outdoor kennels and laboratory dogs held alone but given a 90-minute walk regularly were surprisingly similar physiologically and behaviourally, suggesting that the lives of many owned dogs are similar to those of laboratory dogs. The authors suggested the results be interpreted with caution, but the similarity between the two groups of dogs is disquieting and should stimulate more research into the physiological status of owned dogs that spend much time alone.

Many working dogs live in groups, and this allows them a social life independent of their owners. Their physical life may be harsh, but watching a working dog at work suggests they are enjoying the activity. The welfare of the single dog depends on the willingness of the owner to spend as much quality time with the dog as possible. There are limits to the time available for many owners but these limitations can be countered in several ways, including taking the dog to work, using a dog-minding organisation similar to child-care, or dog-walkers. Nevertheless, the little time humans spend with their single dog is probably the most limiting factor on the welfare of the companion dog. The Blue Cross animal welfare charity in the UK reported that increasingly people relinquish their dogs for re-homing because they do not have enough time. In 2001, 26% of people who gave up their dogs for re-homing did so because they were out all day and had no time for the animal (Anonymous, 2003c).

The physical welfare of pet dogs has certainly improved, in recent times, but the psychological welfare of many may have worsened. It may be that there is no space for dogs in many modern urban societies, and trying to make the latter suited for dogs may result in failure. It may be necessary to modify the dog to be an inactive, non-aggressive, quiet animal, a living plaything for humans. Genetic manipulation and cloning will allow for a better understanding of the genetics of behaviour and these scientific developments may allow for rapid changes in the temperament of dogs to be used as pets. There are of course, welfare problems with cloning, including the health of newborn puppies.

In suburban Melbourne, a study of dogs kept in back yards found that 31% of them were walked once a week or less and of the dogs that were walked a little over half of them had time off the leash (Kobelt *et al.*, 2003b). In the USA, 29% of respondents walked their dog once or twice a week (Slater *et al.*, 1995), and 65% of dogs were allowed off the leash (Table 4). Respondents of surveys such as these may be more dog-orientated than non-responding owners, thus the data on exercising dogs may suggest that on average dogs are exercised more than they really are. Walking a dog on a lead may limit the amount of smelling a dog is allowed, and this appears to be one of the things that dogs enjoy while being walked. Many dogs are permitted very little time to explore the smell at the lamp posts and other vertical surfaces they encounter during their daily walk.

Table 4. The amount and type exercise dogs were allowed in Texas, USA.

Activity	Response	Percentage of owners
Play with dog during week	**Never**	**19**
	<1	10
	1-2	20
	4-6	6
	>6	45
Leash walks during week	<1	9
	1-2	20
	3-7	54
	>7	17
Off-leash walks during week	<1	20
	1-2	16
	3-7	40
	>7	24
Reason for walk	Urinate, defaecate	11
	Exercise	80
	Shows	9
Retrieving/playing with disc		62

Adapted from Slater *et al.* (1995)

In the UK, dog behavioural problems are seasonal, with an increase in canine aggression during the summer months and December, possibly due to increased contact with people during holiday periods. Aggression in the home peaked during winter, which may be due to increased confinement indoors. Separation anxiety peaked in January, May and August, after holidays when owners spent lots of time with their dogs. Fear and phobias peaked in November with bonfire night (Association of Pet Behavour

Counsellors, 1995). This suggests that exercise is seasonal and that during the winter, dogs may be exercised less than during summer.

It is thought that dogs need exercise, but if so how much do they need? The requirements will vary from breed to breed and from dog to dog and be influenced by age, health and temperament. Generalisations about dogs and exercise are probably futile. Exercise is an important factor to reduce obesity, but walking on the leash might not be of much benefit. The provision of dog parks in cities, where dogs can be allowed off the leash, is one way of allowing exercise to dogs living in cities. However, in a legal assessment of dog parks, Hannah (2002) suggested that dogs using such parks should have proof of health to prevent the spread of disease, and rules should govern the removal of faeces. There is a danger that unless properly sited and fenced, these parks may be a nuisance to non-dog owners using adjacent areas when dogs wander off their designated areas. Fighting may also be a problem, but is not as common as might be suspected. There is need for research into the use of dog parks and the behaviour of dogs using these areas.

The shift of dogs from living outside to living indoors may cause problems for dogs. A dog living outside is inadvertently automatically treated as a subordinate animal, but when it is moved indoors its perception of its own status may change and it may be elevated in the social hierarchy. The majority of dogs are obviously able to accommodate this change in living conditions, but some may find it difficult to remain subordinate when treated as being important.

There are specific dangers for the dog in the home, including toxic chemicals, windows and verandahs. The Blue Cross reported that in the UK household accidents were the second most common reason for pups being presented for veterinary treatment. Especially dangerous were falls from balconies/windows, scalds and burns, cleaning products, swallowing objects, heavy objects falling, chemicals in garages, bin-raiding, people (shutting doors, walking on or dropping pups), and electrical wires (Anonymous, 2003a). A smoke detector with a low battery warning device that bleeped or chirped made one dog so anxious, that it destroyed a room before damaging an exterior door and escaping (Nash & Watson, 1994). However, the risk of failure of retrieving dogs during evacuation programmes following earthquakes, fires and flooding were greater in families where the dog was kept outdoors all the time (Heath *et al.*, 2001). In addition tethering dogs outdoors may cause problems. Tethered dogs may be more aggressive than dog in kennels and runs and dogs may be strangled if they accidentally hang themselves up on a chain.

5.   NEUTERING

Surgical de-sexing is recommended for pet dogs not intended to be used for breeding. However, a standard recommendation for good welfare is that animals should be allowed to engage in normal behaviour. Mating or competing for mating rights is normal behaviour and de-sexing prevents this. The prevention of these behaviours is considered significant by some, but when balanced against the misery experienced by unwanted pups de-sexing is generally thought to be of little consequence. Vasectomy or tubal ligation would allow sexual activity without the danger of producing unwanted pups. De-sexing also reduces the occurrence of undesirable sexual behaviour such as hypersexuality directed towards humans (Maarschalkerweerd *et al.*, 1997) and there are other health benefits from de-sexing male and female dogs. The Blue Cross reported that a much smaller percentage of dogs surrendered to their shelters for re-homing were unwanted pups from accidental litters compared with 10 years ago, suggesting that the emphasis on neutering has been successful (Anonymous, 2003c).

The pain caused by de-sexing can be reduced significantly by the use of analgesics (see Chapter 6), but pharmacological and physical means of preventing pregnancy or triggering abortion are also available. Interestingly, in Sweden a small percentage of dogs are de-sexed and yet unwanted dogs are not a particularly great problem. Responsible management of dogs prevents most unwanted pregnancies and a small percentage of irresponsible owners cause most of the problems (Murray, 1993)

6.   TRANSPORT AND BOARDING KENNELS

Dogs are transported everywhere and in many different ways. Many dogs are transported in the passenger seat of cars, in the back of station wagons, or on the back of open trucks. There are obvious dangers when a dog travels in the front passenger seat of a car, as it may interfere with the driver and distract his/her attention, or become a flying mass if the car stops suddenly. Safety-belts are now available for dogs and are recommended for all dogs travelling in passenger seats. Dogs in the back of a station wagon should probably be tied there or held in a cage. Some countries outlaw transporting dogs in the trunk (boot) of cars, and where these are tightly sealed an animal could suffocate. In general dogs should be tied onto the back of pick-ups using a short lead secured to the centre of the cab. Farm dogs often work off pick-ups and four-wheel motorcycles. They should not be tied on when working as they can then jump off when required to work livestock. Individual working or sporting dogs may travel in boxes attached to the back of a car or underneath a truck for transporting livestock. International shipment of dogs is usually by air and not a major welfare problem. However, the importation of dogs into rabies-free countries usually results in them spending months in quarantine kennels, and this may become a welfare problem depending on the management of the facilities and the possibility of separation anxiety.

One of the significant experiences of pet dogs is spending time in boarding kennels. Many pet dogs live in single-dog homes and then are expected to spend days or weeks in noisy, often crowded kennels. There is little documented evidence on the effects of boarding, but dogs often lose weight and develop diarrhoea while boarding and return home looking stressed and weakened. The observation of such developments often makes people unwilling to send their dogs to boarding kennels.

The urinary cortisol:creatinine ratios of Labrador Retrievers moved from home into a military kennel increased significantly initially and then declined over the following 10 days. In dogs that had been habituated to kennels previously the ratio returned to pre-kennel levels by 10 days, but dogs with no previous experience of kennels had ratios that were still significantly higher than pre-kennel values at 10 days and even 10 weeks after entering kennels (Gaines *et al.*, 2003). The dogs with high ratios were ranked as having lower training ability than those with low ratios. As many companion dogs are placed in kennels for spells of 7–14 days and are not trained to them beforehand, they probably suffer from stress similar to that seen in the Labrador Retrievers described above.

7.   DOG ABUSE

As dogs become companions the human-dog relationship changes. Good working dogs are valuable animals, and while they may be abused this is financially counter-productive and foolish. However, companion dogs become part of the family and as such may be exposed to all the positive and negative events that close relationships entail. In many families, there are tensions and these can be expressed in a range of ways. At one extreme there is physical violence and the companion animal may be physically abused. However, even in families where physical abuse does not occur but there are tensions, and dogs may respond to them by developing diarrhoea, gastric upsets and seizures (Cain, 1983).

Pet abuse is not uncommon and non-accidental injuries, including sexual injuries and deliberate injury to gain attention, have been described in the UK by Munro and Thrusfield (2001a,b,c,d), from data gained in a survey of veterinarians. Almost all (91%) of the 404 veterinarians who responded accepted that animal abuse occurred but only 48% had identified it. Most of the abused animals were less than 2 years of age and Staffordshire Bull Terriers and mongrels were particularly at risk, while Labrador Retrievers were under-represented.

Veterinarians become suspicious of abuse when the medical history is inconsistent with the injuries, or there is lack of a history. The type of injury and the occurrence of repetitive injuries or multiple problems in the same household are also causes of concern. The behaviour of the dog or the owner were also important features in identifying battered pets (Munro & Thrusfield, 2001a). Many injuries were skeletal, involving fractured ribs or limbs.

In a survey of animal abuse in the UK, 6% of 448 cases were considered to be examples of sexual abuse. The 21 cases involving dogs (Table 5) included a range of injuries that are similar to those identified in human victims of sexual abuse (Munro & Thrusfield, 2001c).

Table 5. Injuries sustained by dogs (n = 21) suspected as being the subjects of sexual abuse. Case No.

Case No	Injury
1	Vaginal injuries, wife saw husband having sex with dog
2	Lodger pleaded guilty to having sex with dog, no injuries seen
3	Vaginal trauma, haemorrhage, possibly 'raped' by human
4	Refractory vaginitis, suggested cause was human 'interference'
5	No lesions, wife suspected husband 'interfered' with bitch
6	Gross vaginal lesions
7	Haemorrhage from vagina, knife wound deep in vagina
8	Knitting needle penetrated uterine/cervical wall
9	Tear in uterus rostral to cervix, scarring through cervix
10	Candle found in vagina
11	Multiple haemorrhages around vulva and internally
12	Piece of broom handle in vagina
13	Piece of stick in vagina
14	Broomstick in rectum
15	Mucosa around anus damaged
16	Anal ring dilated and man seen abusing dog
17	Cord tied around base of penis
18	Elastic band placed around scrotum
19	Evidence of ligature around scrotum – child admitted to placing elastic band
20	Dog castrated and left bleeding – not claimed
21	Large pararectal wound

Adapted from Munro and Thrusfield (2001c)

The injuring of dogs to gain personal attention sometimes called Munchausen syndrome by proxy was identified in 9/448 suspected non-accidental injuries and seven of these involved dogs (Table 6) (Munro & Thrusfield, 2001d). This syndrome is suspected when physical injuries are intentionally produced or invented by the owner who denies causing them and the injuries diminish when the animal is removed from the owner. Owners with this problem often engage in attention-seeking behaviour, the clinical signs and symptoms of which are often quite abnormal; there may be serial deaths of pets, regular incidents over a number of years, a regular change of veterinarian, and children of the owner may also be frequently brought to a doctor for treatment. Although there were few cases identified in this survey, it may be more common than suspected. However, these are complex cases involving the dog's welfare and the owner's mental health. Client confidentiality may limit initial investigation by the veterinarian.

Table 6. Seven suspected cases of Munchausen by proxy syndrome in dogs.

Case No	Incident(s)
1.	Veterinarian called out three times in one day by owner who was insistent that the neighbour had poisoned the dog. Dog recovered in hospital. Owner convicted for attempted poisoning of his child and evidence that he had poisoned other pets
2.	Owner treated dog for haematuria, lameness, and otitis externa, but veterinarian found it difficult to diagnose problem. Dog reported fitting, very abnormal electrolytes, sent to referral centre and died
3.	Client had broken legs of previous dogs. Known to police but could not prosecute as constantly changed veterinarian
4.	Series of incidents over the years led to suspicion of syndrome
5.	Eight to 12 pets died in unexplained circumstances, e.g. after nail-clipping. Attention-seeking behaviour of owner led to suspicion of syndrome
6.	Veterinarian positive owner had poisoned dog. Dog died and owner triumphant but did not allow post mortem
7.	Puppy with head injuries. Veterinarian requested to call to see other puppies, all with crushed skulls. Owner admitted injuring puppies

Adapted from Munro & Thrusfield (2001d)

There is evidence that violence in the home against children or other adults may be accompanied by violence towards animals, and a statistically significant relationship between child and pet abuse in the same household has been shown (De Viney *et al.*, 1983). In the USA, families suspected of child abuse often had more pets than other families living in the same area, but they also had a large turnover of pets with few more than 2 years of age (Anonymous, 2001b). Many people who abuse pets have poor social skills and may be involved in other anti-social crimes such as arson. They often have been abused themselves. Apparently in the USA, 70% of people who abuse animals will commit a violent crime as adults and 95% of these will be male.

In the UK, the Royal Society for the Protection of Cruelty to Animals noted an increase in violence towards animals in 2002, and there were 57 prison sentences for this compared to 46 in 2001. This could indicate a real increase in violence or improved reporting by the general public and harsher sentencing by the judiciary (Anonymous, 2003b).

8.   HOARDING DOGS

The human interest in animals, particularly pet animals, can become extreme and one expression of this is the pathological collecting or hoarding of animals (Patronek, 1999). Many people collect animals or birds but are careful to provide for them. When the collector is unable to care for their animals they can be classed as a pathological hoarder. Compulsive hoarding of inanimate objects is a recognised psychiatric problem but the hoarding of animals is uncommon.

In the USA, the Hoarding of Animals Research Consortium defines an animal hoarder as one who has accumulated a large number of animals which overwhelms the person's ability to provide even minimal nutrition, sanitation, and veterinary care. Hoarders fail to acknowledge the deteriorating condition on the animals, even disease, starvation and death, and filthy household environment and severe overcrowding. They also do not recognise the negative effect of the collection on their health and the well-being of other household members (Kuehn, 2002).

The welfare of dogs owned by hoarders may be compromised. People who have worked with animal welfare organisations often deal with animal hoarders and describe cats or dogs living in appalling conditions. In one American study of 54 separate incidents of hoarding, Patronek (1999) found that hoarders tended to be elderly, solitary females who collected one or two species. Repeat offending was common. Dogs were involved in 60% of the cases documented, and animals were found in poor condition or dead in 80% of cases. The houses lived in by hoarders were usually cluttered with rubbish, and animal faeces and urine were found everywhere, often on the owner's bed (Table 7).

Table 7. Sanitary rating of animal hoarders' residences (n = 54).

Rating	Condition of household	Number
1	Reasonably clean and tidy	3/54
2	Moderately cluttered, with some garbage but no urine or faeces present in living or food preparation areas	8/54
3	Heavily cluttered with garbage. Unsanitary living and food preparation areas. Odour. Urine and faeces in animal cages	4/54
4	Heavily cluttered with garbage. Unsanitary living and food preparation areas. Strong odour. Fresh faeces or urine in human living quarters	12/54
5	Filthy environment, with profuse faeces and urine in living areas	22/54
6	Unknown	5/54

Adapted from Patronek (1999)

Hoarders justify their behaviour as a love of the animals, and a fear, probably justified, that the animals would be killed if taken to a shelter. They are concerned that no-one else would care for the animals (Patronek, 1999). Although hoarders are usually poor people there are cases of professional people also engaging in hoarding (Kuehn, 2002).

9.   GRIEF AT EUTHANASIA

Euthanasia of pet dogs occurs under three defined circumstances: unwanted dogs at shelters, healthy dogs killed at the behest of their owners, and old dogs ill killed because their quality of life has deteriorated greatly. The euthanasia of a beloved but ill and ageing dog is always sad, but one of the good things about veterinary medicine is that an animal need not be allowed to suffer unnecessarily.

However, more dog owners are now unwilling to have a pet dog killed even when its quality of life is seriously compromised. This may be due to the owner viewing the animal as a child and being willing to make a greater effort and spend more money on keeping it alive (Anonymous, 2004). This may be to the detriment of the animal's welfare. Owners may also be less willing to accept the opinion of a veterinarian and may have greater expectations of veterinary geriatric medicine than is commonly available. It is a difficult situation as the principal concern of the veterinarian must be the dog's welfare. A greater percentage of dogs now live to old age and therefore are subjected to the diseases and conditions of ageing such as cancers, cognitive dysfunction and cardiovascular disease. Such long-term relationships may make it more difficult for owners to have their dog killed, and this is supported by the demand for treatment for cognitive dysfunction and for a diet that slows or reverses this development.

In contrast, a healthy well-behaved dog may be presented for euthanasia. This may be due to the collapse of the human-dog bond or it never having developed. Financial difficulties, or family changes (birth, divorce, bereavement) and lifestyle may precipitate the decision to kill the animal (Anonymous, 2004). Well-behaved healthy dogs may be re-homed, but this depends on the willingness of the owner to allow it to happen.

The elevation of the dog to a family member changes the attitude of the owner towards the dog if it becomes seriously ill or is severely injured. Previously, because veterinary medicine was incapable or unwilling to treat severely injured animals, they were either subject to euthanasia or treated inadequately and left to malinger and often die. Today, with adequate insurance cover, good veterinary medicine and changed expectations of veterinary care and a dog's longevity and health, owners may expect that their dog (family member) can be cured regardless of the process and the pain involved.

10.   CONCLUSIONS

There is much about the welfare of pet dogs which is unknown. The survival to adulthood of pups bred to be pets has not been quantified, nor has the proportion of dogs which die or are killed at various stages of their juvenile lives. That dog numbers are decreasing in some European countries suggests that some owners have decided that they cannot provide the pet dog with what it needs socially and physically. It is ironic that as our knowledge of canine health and nutrition is expanding, an inability to provide what dogs need in other ways may be recognised as a significant deterrent to dog ownership. Although we have not defined exactly what it is a dog requires socially or physically of its environment, many people recognise that they cannot supply a dog with an adequate environment given their modern, busy and mobile lifestyles.

Maintaining a human-dog relationship and providing for the dog may be as difficult for some as maintaining a human-human relationship. The observation that many dogs in large cities are never let off the lead but are constantly restricted, and that such is the future for many dogs in large cities supports the contention that dogs are an anachronism in modern urban societies, both from the dog's perspective and society's. Dogs living in restricted households may be experiencing chronic stress like dogs in laboratory cages.

The anthropocentric concept that people should own dogs because dog ownership is beneficial for humans, without the corollary that the life of a dog as a pet might not be good for the dog, has not been investigated sufficiently. The description of a dog's life in suburban Melbourne by Kobelt (2004) is disturbing. Those dogs lived in back gardens but had nothing to do and were essentially inactive. Dogs living indoors have probably similarly dull lives.

REFERENCES

Abeynayake, P. 1997. Management of pain in dogs and cats using non-steroidal anti-inflammatory analgesics. Australian Veterinary Journal 75, 353.

Adams, G.J. and Clark, W.T. 1989. The prevalence of behavioural problems in domestic dogs; a survey of 105 dog owners. Australian Veterinary Practice 19, 135-7.

Adams, G.J. and Grandage, J. 1989. Digging behaviour in domestic dogs. Australian Veterinary Journal 66, 126.

Adams, G.J. and Johnson, K.G. 1994a. Behavioural responses to barking and other auditory stimuli during night-time sleeping and waking in the domestic dog (*Canis familiaris*). Applied Animal Behaviour Science 39, 151-62.

Adams, G.J. and Johnson, K.G. 1994b. Sleep, work and the effects of shift work in drug detection dogs, *Canis familiaris*. Applied Animal Behaviour Science 41, 115-26.

Aitken, M.M. 1994. Women in the veterinary profession. Veterinary Record 134, 546-51.

Aktas, M., Auguste, D., Lefebve, H.P., Toutain, P.L. and Braun, J.P. 1993. Creatine kinase in the dog: a review. Veterinary Research Communications 17, 353-369.

Albert, A. and Bulcroft, K. 1987. Pets and urban life. Anthrozoos 1, 9-23.

Albert, A. and Bulcroft, K. 1988. Pets, families and the life course. Journal of Marriage and the Family 50, 532-52.

Alderman, C.E. 2003. Electric shock collars and dog training. Veterinary Record 153, 571-2.

Alexander, S.A. and Shane, S.M. 1994. Characteristics of animals adopted from an animal control center whose owners complied with a spaying/neutering program. Journal of American Veterinary Medical Association 205, 472-6.

Allman, J.M. 1998. Evolving Brains. Scientific American Library, New York, USA.

Altom, E.K., Davenport, G.M., Myers, L.J. and Cummins, K.A. 2003. Effect of dietary fat and exercise on odorant-detecting ability of canine athletes. Research in Veterinary Science 75, 149-55.

Anderson, M.A., Constantinescu, G.M., Dee, L.G. and Dee, J.F. 1995a. Fractures and dislocations of the racing greyhound – Part I. The Compendium for Continuing Education of Practicing Veterinarians 17, 779-898.

Anderson, M.A., Constantinescu, G.M., Dee, L.G. and Dee, J.F. 1995b. Fractures and dislocations of the racing greyhound – Part II. The Compendium for Continuing Education of Practicing Veterinarians 17, 899-909.

Anderson, R.S. 1973. Obesity in the dog and cat. The Veterinary Annual 182-6.

Anonymous. 1995. Veterinarians probe greyhound idiosyncrasies. Journal of the American Veterinary Medical Association 206, 1689-93.

Anonymous, 1997. Pet ownership data. Journal of American Veterinary Medical Association, 211, 169-70.

Anonymous. 1998a. The complete dog book. Howell Book House, Foster City, USA.

Anonymous. 1998b. Trends in pet ownership. Veterinary Record 142, 473.

Anonymous. 2001a. Animal abuse and human violence: exploring a connection. Veterinary Record, 148, 326-7.

Anonymous. 2001b. Behavioural problems in dogs and cats. Veterinary Record 149, 36.

Anonymous. 2001c. When should bitches be neutered. Veterinary Record 148, 491-3.

Anonymous. 2003a. Household threats to pets. Veterinary Record 153, 546.

Anonymous. 2003b. Increase in violence towards animals. Veterinary Record 152, 607.

Anonymous. 2003c. Less time for pets. Veterinary Record 153, 3-4.

Anonymous. 2003d. New AAHA position statement opposes ear cropping, tail docking. Journal of the American Veterinary Medical Association 223, 1713.

Anonymous. 2004. Dealing with death in man and animals. Veterinary Record 154, 250-1.

Appleby, D.L., Bradshaw, J.W.S. and Casey, R.A. 2002. Relationship between aggressive and avoidance behaviour by dogs and their experience in the first six months of life. Veterinary Record 150, 434-8.

Arkow, P. 1985. The humane society and the human-companion animal bond: Reflections on the broken bond. Veterinary Clinics of North America: Small Animal Practice 15: 455-65.

Aronson, L. 1999. Animal Behavior Case of the Month. Journal of the American Veterinary Medical Association 215, 22-4.

Askew, H.R. 1996. Treatment of behaviour problems in dogs and cats. Blackwell, London, England.

Association of Pet Behaviour Counsellors. 1995. Report. Veterinary Record, 136, 159.

Balcom, S. and Arluke, A. 2001. Animal adoption a negociated order: A comparison of open versus traditional shelter approaches. Anthrozoos 14, 135-150.

Bandow, J.H. 1996. Will breed-specific legislation reduce dog bites. Canadian Veterinary Journal 37, 478-81.

Bartels, K.E., Stair, E.L. and Cohen, R.E. 1991. Corrosion potential of steel bird shot in dogs. Journal of the American Veterinary Medical Association 199, 856-63.

Baskin, C.R., Hinchcliff, K.W., DiSilvestgro, R.A., Reinhart, G.A., Hayek, M.G., Chew, B.P., Burr, J.R. and Swenson, R.A. 2000. Effects of dietary antioxidant supplementation on oxidative damage and resistance to oxidative damage during prolonged exercise in sled dogs. American Journal of Veterinary Research 61, 886-91.

Bate, M. 1997. The dog as an experimental animal. The Australian and New Zealand Council for the care of Animals in Research and Teaching News 10, 1-8.

Bateson, P. (1991). Assessment of pain in animals. Animal Behaviour 42, 827-39.

Beaudet, R., Chalifoux, A. and Dallaire, A. 1994. Predictive value of activity level and behavioral evaluation on future dominance in puppies. Applied Animal Behaviour Science 40, 273-84.

Beaver, B.V. 1994. Owner complaints about canine behaviour. Journal of the American Veterinary Medical Association 204, 1953-5.

Beaver, B.V. 1997. Human-Canine Interactions: A summary of perspectives. Journla of the American Veterinary Medical Association 210, 1148-50.

Beaver, B.V., Fischer, M. and Atkinson, C.E. 1992. Determination of favourite components of garbage by dogs. Applied Animal Behaviour Science 34, 129-36.

Beaver, B.V., Reed, W., Leary, S., McKiernan, B., Bain, F., Schultz, R., Bennett, B.T., Pascoe, P., Shull, E., Cork, L.C., Francis-Floyd, R., Amass, K.D., Johnson, R., Schmidt,

R.H., Underwood, W., Thornton, G.W. and Kohn, B. 2001. Report of the AVMA Panel on euthanasia. Journal of the American Veterinary Medical Association 218, 669-97.

Bebak, J. and Beck, A.M. 1993. The effect of cage size on play and aggression between dogs in purpose-bred beagles. Laboratory Animal Science 43, 457-9.

Beck, A.M. 1973. The ecology of stray dogs: A study of free-ranging dogs in Baltimore. York Press, Baltimore, USA.

Beck, A.M. 1975. The ecology of "feral" and free-roving dogs in Baltimore. IN: The Wild Canids, edited by MW Fox. Behavioral Science Series. Van Nostrand Reinhold Co, New York, USA.

Beck, A.M. 2000. The human-dog relationship: A tale of two species. IN: Dogs, zoonoses and public health, edited by Macpherson C.N.L., Meslin, F.X. and Wandeler, A.I. CAB International, Wallingford UK. Pp. 1-16.

Beckett, S. 1962. Malone dies. Penguin Books Ltd, Middlesex, England. Pp. 21.

Beer, L. 2003. Recognition of stress behaviour in dogs participating in competitive dog sports. In Proceedings of 4th International Veterinary Behavioural Meeting; Proceedings 352, Post Graduate Foundation in Veterinary Science, University of Sydney, Australia, 251-3.

Beerda, B., Schilder, M.B.H., Janssen, N.S.C.R.M. and Mol, J.A. 1996. The use of saliva cortisol, urinary cortisol and catecholamines measurements for a non-invasive assessment of stress responses in dogs. Hormones and Behaviour 30, 272-9.

Beerda, B., Schilder, M.B.H., van Hoff, J.A.R.A.M. and de Vries, H.W. 1997. Manifestation of chronic and acute distress in dogs. Applied Animal Behavioural Science 52, 307-19.

Beerda, B., Schilder, M.B.H., van Hooff, J.A.R.A.M., de Vries, H.W. and Mol, J.A. 1998. Behavioural, saliva cortisol and heart rate responses to different types of stimuli in dogs. Applied Animal Behaviour Science 58, 365-81.

Beerda, B., Schilder, M.B.H., van Hooff, J.A.R.A.M., de Vries, H.W. and Mol, J.A. 1999a. Chronic stress in dogs subjected to social and spatial restrictions. I. Behavioural responses. Physiology and Behaviour 66, 233-42.

Beerda, B., Schilder, M.B.H., van Hooff, J.A.R.A.M., de Vries, H.W. and Mol, J.A. 1999b. Chronic stress in dogs subjected to social and spatial restrictions. II. Hormonal and immunological response. Physiology and Behaviour 66, 233-42.

Beerda, B., Schilder, M.B.H., van Hooff, J.A.R.A.M., de Vries, H.W. and Mol, J.A. 2000. Behavioural and hormonal indicators of enduring environmental stress in dogs. Animal Welfare 9, 49-62.

Bennett, P.C. and Perini, E. 2003. Tail docking of dogs: can attitude change be achieved. Australian Veterinary Journal 81, 277-82.

Best, L.W., Corbett, L.H., Stephens, D.R. and Newsome, A.E. 1974. Baiting trials for dingoes in central Australia with poison 1080, encapsulated strychnine and strychnine suspended in methyl cellulose. CSIRO Division of Wildlife Research Technical Paper No. 30, Pp.1-7.

Bhanganada, K., Wilde, H., Sakolsataydorn, P and Oonsombat, P. 1993. Dog-bite injuries at a Bangkok teaching hospital. Acta Tropica 55, 249-55.

Bhargava, H.N. (1994). Diversity of agents that modify opioid tolerance, physical dependence, abstinence syndrome and self administration behaviour. Pharmacological Reviews 46, 293-324.

Bird, P. 1994. Improved electric fences and baiting technique – a behavioural approach to integrated dingo control. Final report to the Wool Research and Development Corporation on Project DAS39.

Blair, B.J. and Townsend, T.W. 1983. Dog predation of domestic sheep in Ohio. Journal of Range Management 36, 527-8.

Bloom, P. 2004. Can a dog learn a word. Science 304, 1605-6.

Bloomberg, M.S., Dee, J.F., Taylor, R.A. and Gannon, J.R. 1998. Canine sports medicine and surgery. W.B. Saunders Company, Philadelphia, USA.

Bloomberg, M.S. and Dugger, W.W. 1998. Greyhound racing injuries: Racetrack injury survey. IN Canine sports medicine and surgery. Eds M.S. Bloomberg, J.F. Dee, R.A. Taylor and J.R. Gannon. W.B. Saunders Company, Philadelphia, USA. Pp 412-5.

Boitani, L., Francisci, F., Cuicci, P. and Andreoli, G. 1995. Population biology and ecology of feral dogs in central Italy. IN: The Domestic Dog: its evolution, behaviour and interaction with people, edited by J. Serpell. Cambridge University Press, Cambridge, England. Pp. 217-44.

Boivin, C.P. 1998. Oversupply of veterinarians in Canada could lead to a significant drop in income, study concludes: Demographic analysis of the veterinary profession in Canada commissioned by the CVMA. Canadian Veterinary Journal 39, 535-6.

Bojrab, M.J.M., Crane, S.W. and Arnolzky, S.P. 1983. Current techniques in small animal surgery. Lea and Febiger, Philadelphia, USA.

Bonnett, B.N., Egenvall, A., Olson, P. and Hedhammer, A. 1997. Mortality in insured Swedish dogs: rates and causes of death in various breeds. Veterinary Record 141, 40-4.

Boudrieau, R.J., Dee, J.F. and Dee, L.G. 1984a. General tarsal bone fractures in the racing greyhound; a review of 114 cases. Journal of the American Veterinary Medical Association 184, 1486-91.

Boudrieau, R.J., Dee, J.F. and Dee, L.G. 1984b. Treatment of central tarsal bone fractures in the racing greyhound. Journal of the American Veterinary Medical Association 184, 1492-1500.

Bradshaw, J.W.S., Goodwin, D., Lea, A.M. and Whitehead, S.L. 1996. A survey of the behavioural characteristics of pure-bred dogs in the United Kingdom. Veterinary Record 138, 465-8.

Branson, K.R., Ko, J.C.H., Tranquilli, W.J., Benson, J. and Thurmon, J.C. 1993. Duration of analgesia induced by epidurally administered morphine and medetomidine in dogs. Journal of Veterinary Pharmacology and Therapeutics 16, 369-72.

Breur, G.J., Lust, G. and Todhunter, R.J. 2001. Genetics of canine hip dysplasia and other orthopaedic traits. IN The Genetics of the Dog. Eds A. Ruvinsky and J. Sampson. CABI Publishing, Wallingford, England. Pp. 267-98.

Brock, N. 1995. Treating moderate and severe pain in small animals. Canadian Veterinary Journal 36, 658-60.

Brooks, R. 1990. Survey of the dog population of Zimbabwe and its level of rabies vaccination. Veterinary Record 127, 592-6.

Brooks, M. and Sargan, D.R. 2001. Genetic aspects of disease in dogs. IN: The Genetics of the Dog, edited by A. Ruvinsky and J. Sampson. CABI Publishing, Wallingford, England. Pp. 191-266.

Broom, D.M. (2000). Evolution of pain. IN Pain: its nature and management in man and animals, edited by Lord Soulsby of Swaffham and D Morton. The Royal Society of Medicine Press, London, UK. Pp. 17-25.

Brown, R.M. 1998a. Debate regarding tail docking/ear cropping methods. Journal of the American Veterinary Medical Association 213, 472.

Brown, S.G. 1998b. Unique veterinary problems of coursing dogs. IN: Canine sports medicine and surgery, edited by M.S. Bloomberg, J.F. Dee, R.A. Taylor and J.R. Gannon. W.B. Saunders Company, Philadelphia, USA. Pp. 426-33.

Brown, J.P. and Silverman, J.P. 1999. The current and future market for veterinarians and veterinary medical services in the United States. Journal of the American Veterinary Medical Association 215, 161-83.

Bueler, L.E. 1974. Wild dogs of the world. Constable, London.

Bugg, R.J., Robertson, I.D., Elliot, A.D. and Thomson, R.C.A. 1999. Gastrointestinal parasites of urban dogs in Perth, Western Australia. The Veterinary Journal 157, 295-301.

Burger, I.H. and Thompson, A. 1994. Reading a petfood label. IN: The Waltham Book of Clinical Nutrition of the Dog and Cat, edited by J.M. Wills and K.W. Simpson. Pergamon, Elsevier Science Ltd, Oxford, UK. Pp. 15-24.

Burghardt, W.F. 2003. Behavioral considerations in the management of working dogs. The Veterinary Clinics Small Animal Practice 33, 417-46.

Burkholder, W.J. and Toll, P.W. 2000. Obesity. IN: Small Animal Clinical Nutrition, edited by M.S. Hand, C.D. Thatcher, R.L. Remillard and P. Roudebush, Mark Morris Institute, Missouri, USA. Pp. 401-30.

Burr, J.R., Reinhart, G.A., Swenson, R.A., Swain, S.F., Vaughn, D.M. and Bradley, D.M. 1997. Serum biochemical values in sled dogs before and after competing in long-distance races. Journal of the American Veterinary Medical Association 211, 175-9.

Burrows, C. and Ellison, G. 1989. Textbook of veterinary internal medicine 3rd Ed, edited by S.W. Ettiger. Saunders, Philadelphia, USA. Pp. 1559.

Butcher, R. 1999. Stray dogs – a worldwide problem. Journal of Small Animal Practice 40, 458-9.

Butcher, R. 2000. The implementation of stray dog control programmes – The effect of differences in culture and economy. IN Second World Small Animal Veterinary Association Animal Welfare Symposium – Animal welfare issues in urban animal management. Pp. 3-7.

Butler, J.R.A. 2000. Demography and dog-human relationships of the dog population in Zimbabwean communal lands. Veterinary Record 147, 442-6.

Butler, J.R.A. and du Toit, J.T. 2002. Diet of free-ranging domestic dogs (*Canis familiaris*) in rural Zimbabwe: implications for wild scavengers on the periphery of wildlife reserves. Animal Conservation 5, 29-37.

Butler, J.R.A., du Toit, J.T. and Bingham, J. 2004. Free-ranging domestic dogs (*Canis familiaris*) as predators and prey in rural Zimbabwe: threats of competition and disease to large wild carnivores. Biological Conservation 115, 369-78.

Butterwick, R.F. and Hawthorne, A.J. 1998. Advances in dietary management of obesity in dogs and cats. Journal of Nutrition 128, 2771S-5S.

Cain, A.O. 1983. A study of pets in the family system. IN: New perspectives on our lives with companion animals, edited by A.H. Katcher, and A.M. Beck,. University of Pennsylvania, Philadelphia, USA. Pp. 72-81.

Campbell, W.E. 1986a. The effects of social environment on canine behaviour. Modern Veterinary Practice 67, 113-5.

Campbell, W.E. 1986b. The prevalence of behavioural problems in American dogs. Modern Veterinary Practice 67, 28-31.

Campbell, S.A., Hughes, H.C., Griffin, H.E., Landi, M.S. and Mallon, F.M. 1988. Some effects of limited exercise on purpose-bred beagles. American Journal of Veterinary Research 49, 1298-301.

Capner, C.A., Lascelles, B.D.X. and Waterman-Pearson, A.E. 1999. Current British veterinary attitudes to perioperative analgesia for dogs. Veterinary Record 145, 95-9.

Carr, T. 1979. Caudal adhesion subsequent to tail docking. Canine Practice 6, 63-4.

Carter, N.B. 2003. Border collies prove effective in controlling wildlife at airports. ICAO Journal November, 4-7.

Case, D.B. 1987. Dog ownership: a complex web. Psychological Reports 60, 247-57.

Catanzaro, T.E. 1988. A survey of the question of how well veterinarians are prepared to predict their client human-animal bond. Journal of the American Veterinary Medical Association 192, 1707-11.

Caulkett, N., Read, M., Fowler, D. and Waldner, C. 2003. A comparison of the analgesic effects of butorphanol with those of meloxicam after elective ovariohysterectomy in dogs. Canadian Veterinary Journal 44, 565-70.

Causey, M.K. and Cude, C.A. 1980. Feral dogs and white-tailed deer interactions in Alabama. Journal of Wildlife Management 44, 481-4.

Canadian Council on Animal Care. 1984. Guide to the Care and Use of Experimental Animals. Ottawa, Canada.

Chalifoux, A. and Dalliare, A. 1983. Physiologic and behavioural evaluation of CO euthanasia in adult dogs. American Journal of Veterinary Research 44, 2412-7.

Chambers, J.P., Waterman, A.E. and Livingston, A. (1995). The effects of opioid and alpha2 adrenergic blockade on non-steroidal anti-inflammatory drug analgesia in sheep. Journal of Veterinary Pharmacology and Therapeutics 18, 161-6.

Cheetham, S. 2003. Electric shock collars and dog training. Veterinary Record 152, 691.

Childs, J.E., Robinson, L.E., Sadek, R., Madden, A. Miranda, M.E. and Miranda, N.L. 1998. Density estimates of rural dog populations and an assessment of marking methods during a rabies vaccination campaign in the Philippines. Preventive Veterinary Medicine 33, 207-18.

Christiansen, F.O., Bakken, M. and Braastad, B.O. 2001a. Behavioural differences between three breeds of hunting dogs confronted with domestic sheep. Applied Animal Behaviour Science 72, 115-29.

Christiansen, F.O., Bakken, M. and Braastad, B.O. 2001b. Behavioural changes and aversive conditioning in hunting dogs by the second-year confrontation with domestic sheep. Applied Animal Behaviour Science 72, 131-43.

Clark, J.D., Rager, D.R., Crowell-Davis, S. and Evans, D. L. 1997. Housing and exercise of dogs: Effects on behaviour, immune function and cortisol concentration. Laboratory Animal Science 47, 500-10.

Clark, G.I. and Boyer, W.N. 1993. The effect of dog obedience training and behavioural counselling upon the human-canine relationship. Applied Animal Behaviour Science 37, 147-59.

Clark, J.D., Calpi, J.P. and Armstrong, R.B. 1991. Influence of type of enclosure on exercise fitness of dogs. American Journal of Veterinary Research 52, 1028-8.

Cleaveland, S., Kaare, M., Tiringa, P., Mlengeya, T. and Barrat, J. 2003. A dog rabies vaccination campaign in rural Africa: impact on the incidence of dog rabies and human dog-bite injuries. Vaccine 21, 1965-73.

Clevenger, J. and Kass, P.H. 2003. Determinants of adoption and euthanasia of shelter dogs spayed or neutered in the University of California veterinary student surgery program compared to other shelter dogs. Journal of Veterinary Medical Eduction 30, 372-8.

Clifford, D.H., Boatfield, M.P. and Rubright, J. 1983. Observations on fighting dogs. Journal of the American Veterinary Medical Association 183, 654-7.

Clutton-Brock, J. 1995. Origins of the dog: domestication and early history. IN The Domestic Dog; its evolution, behaviour and interaction with people, edited by J. Serpell. Cambridge University Press, Cambridge, England. Pp. 7-20.

Collins, M.R. 1993. Docking of dogs. Veterinary Record 132, 444.

Constable, P.D., Hinchcliff, K.W., Olson, J.L. and Stepien, R.L. 2000. Effects of endurance training on standard and signal-averaged electrocardiograms of sled dogs. American Journal of Veterinary Research 61, 582-8.

Conzemius, M.G., Hill, C.M., Sammarco, J.L. and Perkowski, S.Z. (1997). Correlations between subjective and objective measures used to determine severity of postoperative pain in dogs. Journal of the American Veterinary Medical Association 210, 1619-22.

Cook, C.J. (1997). Oxytocin and prolactin supress cortisol response to acute stress in both lactating and non lactating sheep. Journal of Dairy Research 64, 327-39.

Cooper, J. and McGreevy, P. 2002. Stereotypic behaviour in the stabled horse: causes, effects and prevention without compromising horse welfare. In: The welfare of the horse, edited by N.Waran,. Kluwer Academic Press, Dordrecht, the Netherlands. Pp. 99-124.

Cooper, J.J., Ashton, C., Bishop, S., West, R., Mills, D.S. and Young, R.J. 2003. Clever hounds: social cognition in the domestic dog (Canis familiaris). Applied Animal Behaviour Science 81, 229-44.

Coppinger, R., Glendenning, J., Torop, E., Matthay, C., Sutherland, M. and Smith, C. 1987. Degree of behavioural neoteny differentiates canid polymorphs. Ethology 75, 89-108.

Corbett, L.K. 1995. The dingo in Australia and Asia. Cornell University Press, Ithaca, New York, USA.

Corbett, L.K. and Newsome, A.E. 1975. Dingo society and its maintenance: a preliminary analysis. IN The wild canids, their systematics, behavioural ecology and evolution, edited by M.W. Fox,. Van Nostrand Reinhold, New York, USA. Pp. 366-79.

Craig, A.M. 1998. Drugs and medications: Quality Assurance/Testing programs. IN Canine sports medicine and surgery, edited by M.S. Bloomberg, J.F. Dee, R.A. Taylor and J.R. Gannon. W.B. Saunders Company, Philadelphia, USA. Pp. 383-7.

Crane, S.W., Griffin, R.W. and Messent, P.R. 2000. Introduction to commercial pet foods. IN Small Animal Clinical Nutrition, edited by M.S. Hand, C.D. Thatcher, R.L. Remillard and P. Roudebush,. Mark Morris Institute, Topeka, USA. Pp. 111-26.

Creel, S. and Creel, N.M. 2002. The African Wild Dog. Monographs in Behavior and Ecology, Princeton University Press, Princeton, USA.

Crenshaw, W.E. and Carter, C.N. 1995. Should dogs in animal shelters be neutered early? Veterinary Medicine 90, 756-60.

Cronin, G.M., Hemsworth, P.H., Barnett, J.L., Jongman, E.C., Newman, E.A. and McCauley, I. 2003. An anti-barking muzzle for dogs and its short-term effects on behaviour and saliva cortisol concentrations. Applied Animal Behaviour Science 83, 215-26.

Crowell-Davis, S.L., Barry, K., Ballam, J.M. and Laflamme, D.P. 1995. The effect of caloric restriction of pen-housed dogs: Transition from unrestricted to restricted diet. Applied Animal Behaviour Science 43, 27-41.

Crowell-Davis, S.L., Seibert, L.M., Sung, W., Parthasarathy, V. and Curtis, T.M. 2003. Use of clomipramine, alprazolam and behaviour modification for treatment of storm phobia in dogs. Journal of the American Veterinary Medical Association 222, 744-8.

Cuddy, D. and Dalton, P. 1998. Whelping, raising and training of racing greyhounds. IN Canine sports medicine and surgery, edited by M.S. Bloomberg, J.F. Dee, R.A. Taylor and J.R. Gannon,. W.B. Saunders Company, Philadelphia, USA. Pp. 366-8.

Cui, J. and Wang, Z.Q. 2001. Outbreaks of human trichinellosis caused by consumption of dog meat in China. Parasite – Journal de la Societe Francaise de Parasitologie 8, S74-7.

Dahlbom, M., Andersson, M., Huszenicza, G. and Alanko, M. 1995. Poor semen quality in Irish wolfhounds: a clinical, hormonal and spermatological study. Journal of Small Animal Practice 36, 547-52.

Danbury, T.C., Weeks, C.A., Chambers, J.P., Waterman-Pearson, A.E. and Kestin, S.C. (2000). Self-selection of the analgesic drug carprofen by lame broiler chickens. The Veterinary Record 146, 307-11.

Daniels, T.J. 1983a. The social organisation of free-ranging urban dogs. I. Non-estrus social behaviour. Applied Animal Ethology 10, 341-63.

Daniels, T.J. 1983b. The social organisation of free-ranging urban dogs. II. Estrus groups and the mating system. Applied Animal Ethology 10, 365-73.

Daniels, T.J. and Bekoff, M. 1989a. Spatial and temporal resource use by feral and abandoned dogs. Ethology 81, 300-12.

Daniels, T.J. and Bekoff, M. 1989b. Population and social biology of free-ranging dogs, Canis familiaris. Journal of Mammalogy 70, 754-62.

Darke, P.G.G., Thrusfield, M.V. and Aitken, C.G.G. 1985. Association between tail injuries and docking in dogs. Veterinary Record 116, 409.

Darvelid, A.W. and Linde-Forsberg, C., 1994. Dystocia in the bitch; a retrospective study of 182 cases. Journal of Small Animal Practice 35, 402-7.

David, H. and Balfour, A.D. 1992. The inevitable bond. IN The inevitable bond – Examining Scientist-Animal Interaction, edited by H. David, and A.D. Balfour,. Cambridge University Press, Cambridge, UK. Pp. 27-43.

Davis, S.J.M. and Valla, F.R. 1978. Evidence for domestication of the dog 12,000 years ago in the Natufian of Israel. Nature 276, 608-10.

Davis, M.S., Willard, M.D., Nelson, S.E., McCullough, S.M., Mandsager, R.E., Roberts, J. and Payton, M.E. 2003a. Efficacy of omeprazole for the prevention of exercise-induced gastritis in racing Alaskan sled dogs. Journal of Veterinary Internal Medicine 17, 163-6.

Davis, M.S., Willard, M.D., Nelson, W.S.L., Mandsager, R.E., McKiernan, B.S., Mansett, J.K. and Lehenbauer, T.W. 2003b. Prevalence of gastric lesions in racing Alaskan sled dogs. Journal of Veterinary Internal Medicine 17, 311-4.

De Jong, P. 1992. Justification of the OE: Canine Care. Vetscript, V, 10-12.

De Viney, E., Dickert, J. and Lockwood, R. 1983. The care of pets within child abusing families. International Journal for the study of Animal Problems 4, 321-9.

Dean, S. 1990. Tail docking. Veterinary Record 126, 296.

Debraekeleer, J. 2000. Canine body weights and height at withers. IN Small Animal Clinical Nutrition, edited by M.S. Hand, C.D. Thatcher, R.L. Remillard, and P. Roudebush,. Mark Morris Institute, Topeka, USA. Pp. 1037-46.

Dee, L.G. 1998. Racing Greyhound adoption programmes. IN Canine sports medicine and surgery, edited by M.S. Bloomberg, J.F. Dee, R.A. Taylor and J.R. Gannon,. W.B.Saunders Company, Philadelphia, USA. Pp. 447-9.

de Napoli, J.S., Dodman, N.H., Shuster, L., Rand, W.R. and Gross, K.L. 2000. Effect of dietary protein content and tryptophan supplementation on dominance aggression, territorial aggression and hyperactivity in dogs. Journal of the American Veterinary Medical Association 217, 504-8.

deVile, C.P. 1998. Early neutering. Veterinary Record 142, 228.

Di Lampedusa, G. 1963. The Leopard. Fontana Books, London, UK. Pp. 202.

Diamond, J. 2002. Evolution, consequences and future of plant and animal domestication. Nature 418, 700-7.

DiGiacomo, N., Arluke, A. and Patronek, G. 1998. Surrendering pets to shelters: The relinquisher's perspective. Anthrozoos 11, 41-51.

Dinniss, A.S., Stafford, K.J., Mellor, D.J., Bruce, R.A. and Ward, R.N. 1997. Acute cortisol responses of lambs castrated and tailed using rubber rings with or without a castrating clamp. Australian Veterinary Journal 75, 494-7.

Dobromylskyj, P., Flecknell, P.A., Lacelles, B.D., Livingston, A., Taylor, P. and Waterman-Pearson, A. 2000. Management of postoperative pain and other acute pain. IN Pain management in animals, edited by P. Flecknell, and A. Waterman-Pearson,. W.B. Saunders, London, UK. Pp. 53-79.

Dobson, G.P., Parkhouse, W.S., Weber, J.M., Stuttard, E., Harman, J., Snow, D.H. and Hochachka, P.W. 1988. Metabolic changes in skeletal muscle and blood of greyhounds during 800 m track sprint. American Journal of Physiology 255, R513-9.

Dodman, N.H., Reisner, L., Shuster, L., Rand, W., Luescher, A., Robinson, I. and Houpt, K.A. 1996. Effect of dietary protein content on behaviour in dogs. Journal of the American Veterinary Medical Association 208, 376-9.

Dohoo, S.E. and Dohoo, I.R. 1996. Postoperative use of analgesics in dogs and cats by Canadian veterinarians. Canadian Veterinary Journal 37, 546-51.

Donnelley, S. 1990. Animals in science: The justification issue. IN Animals Science and Ethics, edited by S. Donnelley and K. Nolan. The Hastings Centre, Garrison, New York, USA. Pp. 8-13.

Donoghue, S and Kronfeld, D.S. 1994. Feeding hospitalised dogs and cats. In The Waltham Book of Clinical Nutrition for the dog and cat, edited by J.M. Wills and K.W. Simpson. Pergamon Elsevier, Oxford, UK. Pp. 25-38.

Douglas, K. 2000. Mind of a dog. New Scientist 165, 24-7.

Douglas, K. 2004. It's good to bark. New Scientist 182, 52-2.

Drickamer, L.C. and Vessey, S.H. 1982. Animal Behaviour: Concepts. Processes and Methods. Willard Grant Press, Boston, USA.

Duhaime, R.A., Norden, D., Corso, B., Mallonee, S. and Salman, M.D. 1998. Disaster medicine- Injuries and illnesses in working dogs used during the disaster response after the bombing in Oklahoma City. Journal of the American Veterinary Medical Association 212, 1202-7.

Dumonceaux, G.A. and Beasley, V.R. 1990. Emergency treatments for police dogs used for illicit drug detection. Journal of the American Veterinary Medical Association 197, 185-7.

Duncan, I.J.H., Rushen, J. and Lawrence, A.B. 1993. Conclusions and implications for animal welfare. IN Stereotypic Animal Behaviour: Fundamentals and Applications to Welfare, edited by A.B. Lawrence. CABI Wallingford, UK. Pp. 193-206.

Dunlop, R.H. and Williams, D.J. 1996. Veterinary Medicine – an illustrated history. Mosby, St. Louis, USA.

Duxbury, M.M., Jackson, J.A., Line, S.W. and Anderson, R.K. 2003. Evaluation of association between retention in the home and attendance at puppy socialization classes. Journal of the American Veterinary Medical Association 223, 61-6.

Dysko, R.C., Nekzek, J.A., Levin, S.I., DeMarco, G.J. and Moalli, M.R. 2002. Biology and Diseases of Dogs. IN Laboratory Animal Medicine, edited by J.G. Fox, L.C Anderson, F.M. Loew and F.W. Quimby. Academic Press, New York, USA.

Eaton, R.L. 1969. Cooperative hunting by cheetahs and jackals and a theory of domestication of the dog. Mammalia 33, 87-92.

Edney, A.T.B. 1998. Reasons for the euthanasia of dogs and cats. Veterinary Record 143, 114.

Edney, A.T.B. and Smith, P.M. 1986. Study of obesity in dogs visiting veterinary practices in the United Kingdom. Veterinary Record 118, 301-6.

Edwards, D. 2004. Vaccination: a victim of its own success. Journal of Small Animal Practice 45, 535.

Egenvall, A., Bonnett, B.N., Olson, P. and Hedhammer, A. 2000a. Gender, age, breed and distribution of morbidity and mortality in insured dogs in Sweden during 1995 and 1996. Veterinary Record 146, 519-25.

Egenvall, A., Bonnett, B.N., Olson, P. and Hedhammer, A. 2000b. Gender, age and breed pattern for veterinary care in insured dogs in Sweden during 1996. Veterinary Record 146, 551-7.

Egenvall, A., Bonnett, B.N., Shoukri, M., Olson, P., Hedhammer, A. and Dohoo, I. 2000c. Age pattern of mortality in eight breeds of insured dogs in Sweden. Preventive Veterinary Medicine 46, 1-14.

Elliot, O. and Scott, J.P. 1961. The development of emotional distress reactions to separation in puppies. Journal of Genetic Psychology 99, 3-32.

Endenburg, N. and Knol, B.W. 1994. Behavioural, household, and social problems associated with companion animals: opinions of owners and non-owners. The Veterinary Quarterly 16, 130-4.

Engeman, R.M., Vice, D.S., Tork, D. and Gruver, K.S. 2002. Sustained evaluation of the effectiveness of detector dogs for locating brown tree snakes in cargo outbound from Guam. International Biodeterioration and Biodegradation 49, 101-6.

Evans, J.M. and Sutton, D.J. 1989. The use of hormones especially progestagens to control oestrus in bitches. IN Dog and cat reproduction, contraception and artificial insemination., editged by P.W. Concannon, D.B. Morton and B.J. Weir. Journal of Reproduction and Fertility; Supplement 39, 163-73.

Evason, M.D., Carr, A.P., Taylor, S.M. and Waldner, C.L. 2004. Alterations in thyroid hormone concentrations in healthy sled dogs before and after athletic conditioning. American Journal of Veterinary Research 65, 333-7.

Ewer, R.F. 1973. The Carnivores. Weidenfeld and Nicolson, London, UK.

Eze, C.A. and Eze, M.C. 2002. Castration, other management practices and socio-economic implications for dog keepers in Nsukka area, Enuga State, Nigeria. Preventive Veterinary Medicine 55, 273-80.

Fayrer- Hosken, R.A., Dookwah, H.D., Brandon, C.I., Forsberg, M., Greve, T., Gustaffson, H., Katila, T., Kindahl, H. and Ropstad, E. 2000. Immunocontrol in dogs. Animal Reproduction Science 60/61, 365-73.

Ferrari, J.R., Loftus, M.M. and Pesek, J. 1999. Young and older caregivers at homeless animal and human shelters: Selfish and selfless motives in helping others. Journal of social distress and the homeless 8, 37-49.

Fifield, S.J. and Forsyth, D.K. 1999. A pet for the children: factors related to family pet ownership. Anthrozoos 12, 24-32.

Firth, A.M. and Haldane, S.L. 1999. Development of a scale to evaluate postoperative pain in dogs. Journal of the American Veterinary Medical Association 215, 651-9.

Flannigan, G. and Dodman, N.H. 2001. Risk factors and behaviours associated with separation anxiety in dogs. Journal of the American Veterinary Medical Association 219, 460-6.

Flecknell, P. and Avril Waterman-Pearson, 2000. Pain management in animals. W.B. Saunders, London, UK.

Flecknell, P. (2000). Recognition and assessment of pain in animals. INnPain: its nature and management in animals, edited by Lord Soulsby of Swaffham and D. Morton. The Royal Society of Medicine Press, London, UK. Pp. 63-8.

Flecknell, P.A. and Molony V. 1997. Pain and Injury. IN Animal Welfare, edited by M.C. Appleby and B.O. Hughes. CABI Publishing, Wallingford, England. Pp. 63-73.

Fleming, P.J.S. 1996. Ground-placed baits for the control of wild dogs: Evaluation of a replacement-baiting strategy in North-eastern New South Wales. Wildlife Research 23, 729-40.

Fleming, P.J.S. and Korn, T.J. 1989. Predation of livestock by wild dogs in eastern New South Wales. Australian Rangelands Journal 11, 61-6.

Fleming, P.J.S., Thompson, J.A. and Nicol, H.I. 1996. Indices for measuring the efficacy of aerial baiting for wild dogs control in North-eastern New South Wales. Wildlife Research 23, 665-674.

Flower, P.J. 1991. Control of fighting dogs. Veterinary Record 128, 553.

Font, E. 1987. Spatial and social organisation: stray urban dogs revisited. Applied Animal Behaviour Science 17, 319-28.

Fowler, D., Isakow, K., Caulgett, N. and Waldner, C. 2003. An evaluation of the analgesic effects of meloxicam in addition to epidural morphine/mepivacaine in dogs undergoing cranial cruciate ligament repair. Canadian Veterinary Journal 44, 643-8.

Fowler, K.J., Sahhar, M.A. and Tassicker, R.J. 2000. Genetic counselling for cat and dog owners and breeders – managing the emotional impact. Journal of the American Veterinary Medical Association 216, 498-501.

Fox, M.W. 1986. Laboratory animal husbandry: Etholoy, welfare and experimental variables. Albany State University of New York Press, New York, USA.

Fox, S.M. 1999. Painful decisions: When and how to treat pain. Irish Veterinary Journal 52, 553-62.

Fox, S.M., Mellor, D.J., Firth, E.C., Hodge, H. and Lowoko, C.R.O. 1994. Changes in plasma cortisol concentrations before, during and after analgesia and anaesthesia plus ovariohysterectomy in bitches. Research in Veterinary Science 57, 110-8.

Fox, S.M., Mellor, D.J., Lowoko, C.R.O., Hodge, H. and Firth, E.C. 1998. Changes in plasma cortisol concentrations in response to different combinations of halothane and butorphanol with or without ovariohysterectomy. Research in Veterinary Science 63, 125-33.

Fox, S.M., Mellor, D.J., Stafford, K.J., Lowoko, C.R.O. and Hodge, H. 2000. The effects of ovariohysterectomy plus different combinations of halothane anaesthesia and butorphanol analgesia on behaviour in the bitch. Research in Veterinary Science 68, 265-74.

Frank, H. and Frank, M.G. 1982. On the effects of domestication on canine social development and behaviour. Applied Animal Ethology 8, 507-25.

Freeman, L.M. and Michel, K.E. 2001. Evaluation of raw food diets for dogs. Journal of the American Veterinary Medical Association 218, 705-9.

Frendin, J. 1998. Chronic suppurative and pyogranulomatous disease ('Stovarsjuka') in hunting dogs. Veterinary Quarterly 20, 77.

Friedmann, E., Katcher, A.H., Lynch, S.A. and Thomas, S.A. 1980. Animal companions and one year survival of patients after discharge from a coronary care unit. Public Health Reports 95, 307-12.

Frommer, S.S. and Arluke, A. 1999. Loving them to death: Blame-displacing strategies of animal shelter workers and surrenders. Society and Animals 7, 1-16.

Gaines, S.A., Rooney, N.J. and Bradshaw, J.W.S. 2003. Physiological and Behavioural responses of dogs to kennelling. In Proceeding of 4th International Veterinary Behavioural Meeting. Proceedings No 352 of the Post Graduate Foundation in Veterinary Science, University of Sydney. Pp. 231.

Galac, S. and Knol, B.W. 1997. Fear-motivated aggression in dogs: Patient characteristics, diagnosis and therapy. Animal Welfare 6, 9-15.

Gannon, J.R. 1998. Drug control programs in canine sports medicine. IN Canine sports medicine and surgery, edited by M.S. Bloomberg, J.F. Dee, R.A. Taylor and J.R. Gannon. W.B.Saunders Company, Philadelphia, USA. Pp. 388-90.

Gavgani, A.S.M., Mohite, H., Edrissian, G.H., Mohebale, M. and Davies, C.R. 2002a. Domestic dog ownership in Iran is a risk factor for human infection with Leishmania infantum. American Journal of Tropical Medicine and Hygeine 67, 511-5.

Gavgani, A.S.M., Hodjati, M.H., Mohite, H. and Davies, C.R. 2002b. Effect of insecticide impregnated dog collars on incidence of zoonotic visceral leishmaniasis in Iranian children. The Lancet 360, 374-9.

Gay. 1984. See lab animal medicine by Fox et al.

Gazit, I. and Terkel, J. 2003. Explosive detection by sniffer dogs following strenuous physical activity. Applied Animal Behaviour Science 81, 149-61.

Goddard, M.E. and Beilharz, R.G. 1982. Genetic and environmental factors affecting the suitability of dogs as guide dogs for the blind. Theoretical and Applied genetics 62, 97-102.

Goodwin, B.K. and Corral, L.R. 1996. Better handicapping and market efficiency in greyhound parimutual gambling. Applied Economics 28, 1181-90.

Goodwin, D., Bradshaw, J.W.S. and Wickens, S.M. 1997. Paedomorphosis affects agonistic visual signals of domestic dogs. Animal Behaviour 53, 297-304.

Grandjean, D., Hinchcliff, K.W., Nelson, S., Schmidt, K.A., Constable, P.D., Sept, R. and Townshend, A.S. 1998. Veterinary problems of racing sled dogs. IN Canine sports medicine and surgery, edited by M.S. Bloomberg, J.F. Dee, R.A. Taylor and J.R. Gannon. W.B. Saunders Company, Philadelphia, USA. Pp. 415-26.

Grandjean, D. 1998. Origin and history of the sled dog. IN Canine sports medicine and surgery, edited by M.S. Bloomberg, J.F. Dee, R.A. Taylor and J.R. Gannon. W.B. Saunders Company, Philadelphia, USA. Pp. 5-9.

Green, J.S. and Woodruff, R.A. 1988. Breed comparisons and characteristics of livestock guarding dogs. Journal of Range Management 41, 249-51.

Gregory, S.P. 1994. Developments in the understanding of the pathophysiology of urethral sphincter mechanism incompetence in the bitch. British Veterinary Journal 150, 135-50.

Grisneaux, E., Pibarot, P., Dupui, J. and Blais, D. 1999. Comparison of ketoprofen and carprofen administered prior to orthopaedic surgery for control of postoperative pain in dogs. Journal of the American Veterinary Medical Association 215, 1105-10.

Grogan, D.P. and Buchholz, R.W. 1981. Acute lead intoxication from a bullet in an intervertebral space. Journal of Bone and Joint Surgery 7, 1180-2.

Gross, W.B. and Siegel, P.B. 1979. Adaptation of chickens to their handlers and experimental results. Avian Diseasees 23, 708-14.

Gross, T.L. and Carr, S.H. 1990. Amputation neuroma of docked tails in dogs. Veterinary Pathology 27, 61-2.

Guccione, G. 1998. Origin and history of the racing greyhound and coursing dogs. IN Canine sports medicine and surgery, edited by M.S. Bloomberg, J.F. Dee, R.A. Taylor and J.R. Gannon. W.B. Saunders Company, Philadelphia, USA. Pp. 1-4.

Guilliard, M.J. 2000. Fractures of the central tarsal bone in eight racing greyhounds. Veterinary Record 147, 512-5.

Gunn, H.M. 1978. The proportions of muscle, bone and fat in two different types of dogs. Research in Veterinary Science 24, 277-82.

Gwaltney-Brant, S.M., Murphy, L.A., Wismer, T.A., and Albretsen, J.C. 2003. General toxicological hazards and risks for search-and-rescue dogs responding to urban disasters. Journal of the American Veterinary Medical Association 222, 292-5.

Hamlin, R.L., Bednarski, L.S., Schuler, C.J., Weldy, P.L. and Cohen, R.B. 1988. Method of objective assessment of analgesia in the dog. Pharmacology and Therapeutics 11, 215-20.

Hand, M.S., Armstrong, P.J. and Allen, T.A. 1989. Obesity: occurrence, treatment and prevention. Veterinary Clinics of North America; Small Animal Practice 19, 447-74.

Hannah, H.W. 2002. Dog parks – are they a public liability? Journal of the American Veterinary Medical Association 220, 312.

Hansen, B.D. 1994. Analgesic therapy. Compendium on continuing veterinary education for practicing veterinarians 16, 868-75.

Hansen, B.D. 2003. Assessment of pain in dogs: Veterinary Clinical Studies. ILAR Journal 44, 197-205.

Hansen, B.D. and Hardie, E. 1993. Prescription and use of analgesics in dogs and cats in a veterinary teaching hospital: 258 cases (1983-1989). Journal of the American Veterinary Medical Association 202, 1485-94.

Hansen, B.D., Hardie, E.M. and Carroll, G.S. 1997. Physiological measurement after ovariohysterectomy in the dog: What's normal? Applied Animal Behaviour Science 51, 101-9.

Hansen, I., Bakken, M. and Braastad, B.O. 1997. Failure of LiCl-conditioned aversion to prevent dogs from attacking sheep. Applied Animal Behaviour Science 54, 251-6.

Hansen, I., Staaland, T. and Ringso, A. 2002. Patrolling with livestock guard dogs: A potentialmethod to reduce predation on sheep. Acta Agriculturae Scandinavica – Section A Animal Science 52, 43-8.

Hardie, E.M. 1996. Chronic pain. IN Predictable Pain Management: A symposium of the North American Veterinary Conference, Orlando, Florida, USA. Pp. 21-6.

Hardie, E.M., Hansen, B.D. and Carroll, G.S. 1997. Behaviour after ovariohysterectomy in the dog. Whats normal? Applied Animal Behaviour Science 51, 111-28.

Hare, B. and Tomasello, M. 1999. Domestic dogs (Canis familiaris) use human and consecific social cues to locate hidden food. Journal of Comparative Psychology 113, 1-5.

Hare, B., Brown, M., Williamson, C. and Tomasello, M. 2002. The domestication of social cognition in dogs. Science 298, 1634-6.

Hart, B.L. 1991. The behaviour of sick animals. Veterinary Clinics of North America: Small Animal Practice 21, 225-37.

Hart, B.L. 1987. Early neutering of pups and kittens. Journal of the American Veterinary medical Association 191, 1181-2.

Hart, B.L. 2001. Effect of gonadectomy on subsequent development of age-related cognitive impairment in dogs. Journal of the American veterinary Medical Association 219, 51-6.

Hart, B.L. and Hart, L.A. 1985a. Selecting pet dogs on the basis of cluster analysis of breed behaviour profile and gender. Journal of the American Veterinary Medical Association 186, 1181-5.

Hart, B.L. and Hart, L.A. 1985b. Canine and Feline Behavioral Therapy. Lea & Febiger, Malvern, USA.

Hart, L.A. 1995. Dogs as human companions: a review of the relationship. IN The Domestic Dog, its evolution, behaviour and interaction with people, edited by J. Serpell. Cambridge University Press, Cambridge, England. Pp. 161-78.

Hart, L.A., Takayanagi, T. and Yamaguchi, H. 1998. Dogs and cats in animal shelters in Japan. Anthrozoos 11, 157-63.

Harvey, C.E., Shofer, F.S. and Laster, L. 1996. Correlation of diet, other chewing activities and periodontal disease in North American client-owned dogs. Journal of Veterinary Dentistry 13, 101-5.

Haug, L.I., Beaver, B.V. and Longnecker, M.T. 2002. Comparison of dogs' reactions to four different head collars. Applied Animal Behavioural Science 79, 53-61.

Haupt, W. 1999. Rabies – risk of exposure and current trends in prevention of human cases. Vaccine 17, 1742-9.

Heath, S.E., Beck, A.M., Kass, P.H. and Glickman, L.T. 2001. Risk factors for pet evacuation failure after a slow-onset disaster. Journal of the American Veterinary Medical Association 218, 1905-10.

Hecht, J. 2002. When did dogs become our best friends? The New Scientist 17, 15.

Hellyer, P.W. 1999a. Minimising postoperative discomfort in dogs and cats. Veterinary Medicine 94, 259-66.

Hellyer, P.W. 1999b. Pain management: Part 1. Veterinary Medicine 94, 253.

Hellyer, P.W. 2002. Treatment of pain in dogs and cats. Journal of the American Veterinary Medical Association 221, 212-5.

Hellyer, P.W. and Gaynor, J.S. 1998. How I treat acute postsurgical pain in dogs and cats. Compendium on continuing education for the practicing veterinarian 20, 140-153.

Hemmer, H. 1990. Domestication: the decline of environmental appreciation. Cambridge University Press, Cambridge, England.

Hemsworth, P.H.J. and Coleman, G.J. 1998. Human-Livestock Interactions. CABI, Wallingford, England.

Hennessy, M.B., David, H.N., Williams, M.T., Mellott, C. and Douglas, C.W. 1997. Plasma cortisol levels of dogs at a county animal shelter. Physiology and Behaviour 62, 485-90.

Hennessy, M.B., Williams, M.T., Miller, D.D., Douglas, C.W. and Voith, V.L. 1998. Influence of male and female petters on plasma cortisol and behaviour: can human interaction reduce the stress of dogs in a public animal shelter. Applied Animal Behaviour Science 61, 63-77.

Hennessy, M.B., Voith, V.L., Mazzei, S.J., Buttram, J., Miller, D.D. and Linden, F. 2001. Behaviour and cortisol levels in dogs in a public animal shelter, and an exploration of the ability of these measures to predict problem behaviour after adoption. Applied Animal Behaviour Science 73, 217-33.

Hennessy, M.B., Voith, V.L., Hawke, J.L., Young, T.L., Centrone, J., McDowell, A.L., Linden, F. and Davenport, G.M. 2002. Effect of a program of human interaction and alterations in diet composition on activity of the hypothalamic-pituitary-adrenal axis in dogs housed in a public animal shelter. Journal of the American Veterinary Medical Association 221, 65-71.

Herbold, J.R., Moore, G.E., Gosch, T.L. and Bell, B.S. 2002. Relationship between incidence of gastric dilatation-volvulus and biometeorologic events in a population of military working dogs. American Journal of Veterinary Research 63, 47-52.

Herin, R.A., Hall, P. and Fitch, J.W. 1978. Nitrogen inhalation as a method of euthanasia in dogs. American Journal of Veterinary Research 39, 989-91.

Herrtage, M.E., Seymour, C.A., White, R.A.S., Small, G.M. and Wight, D.C.D. 1987. Copper toxicosis in the Bedlington terrier, the prevalence in asymptomatic dogs. Journal of Small Animal Practice 28, 1141-51.

Herzog, H.A. and Elias, S.M. 2004. Effects of winning the Westminster Kennel Club dog show on breed popularity. Journal of the American Veterinary Medical Association 255, 365-7.

Herzog, H.A., Betchart, N.S. and Pittman, R.B. 1991. Gender, sex role orientation and attitudes towards animals. Anthrozoos 4, 184-91.

Hetts, S., Clark, J.D., Calpin, J.P., Arnold, C.E. and Mateo, J.M. 1992. Influence of housing conditions on beagle behaviour. Applied Animal Behaviour Science 34, 137-55.

Hewison, C. 1997. Frozen tail or limber tail in working dogs. Veterinary Record 140, 536.

Hewson, C.J. and Luescher, U.A. 1996. Compulsive disorders in dogs. IN Readings in companion animal behaviour, edited by V.L. Voith and P.L. Borchelt. Veterinary Learning Systems, Trenton, USA. Pp. 153-8.

Hiby, E.F., Rooney, N.J. and Bradshaw, J.W.S. 2004. Dog training methods: their use, effectiveness and interaction with behaviour and welfare. Animal Welfare 13, 63-9.

Hickman, J. 1975. Greyhound injuries. Journal of Small Animal Practice 16, 455-60.

Hill, F.W.G. 1985. A survey of the animal population in four high density suburbs in Harare. Zimbabwe Veterinary Journal 16, 31-6.

Hinchcliff, K.W. 1996. Performance failure in Alaskan sled dogs: biochemical correlates. Research in Veterinary Science 61, 271-2.

Hinchcliff, K.W., Reinhart, G.A., Burr, J.R., Schreier, C.J. and Swenson, R.A. 1997a. Metabolizable energy intake and sustained energy expenditure of Alaskan sled dogs during heavy exercise in the cold. American Journal of Veterinary Research 58, 1457-62.

Hinchcliff, K.W., Reinhart, G.A., Burr, J.R., Schreier, C.J. and Swenson, R.A. 1997b. Effect of racing on serum sodium and potassium and acid-base status of Alaskan sled dogs. Journal of the American Veterinary Medical Association 210, 1615-8.

Hinchcliff, K.W., Shaw, L.C., Vukich, N.S. and Schmidt, K.E. 1998. Effects of distance travelled and speed of racing on body weight and serum enzyme activity of sled dogs competing in a long-distance race. Journal of the American Veterinary Medical Association 213, 639-44.

Hinchcliff, K.W., Reinhart, G.A., DiSilvestro, R., Reynolds, A., Blostein-Fujii, A. and Swenson, R.A. 2000. Oxidant stress in sled dogs subjected to repetitive endurance exercise. American Journal of Veterinary Research 61, 512-7.

Hite, M., Hanson, M.H., Bohidar, N.R., Conti, P.A. and Mattis P.A. 1977. Effect of cage size on patterns of activity and health of Beagle dogs. Laboratory Animal Science 27, 60-4.

Hnatkiwskyj, Y.S. 1985. Tending 2 bitches injured during dog fighting. Veterinary Record 117, 190.

Holloway, S.A. 1998. Stress and performance related illness in sporting dogs. IN Canine sports medicine and surgery, edited by M.S. Bloomberg, J.F. Dee, R.A. Taylor and J.R. Gannon. W.B. Saunders Company, Philadelphia, USA. Pp. 28-34.

Holt, P.E. 2000. Early neutering of dogs. Veterinary Record 147, 667.

Holt, P.E. and Thrusfield, M.V. 1993. Association in bitches between breed, size, neutering and docking and acquired urinary incontinence due to incompetence of the urethral sphincter mechanism. Veterinary Record 133, 177-80.

Holt, P.E. and Thrusfield, M.V. 1997. Tail docking in dogs. Australian Veterinary Journal 75, 449.

Holton, L.L., Reid, J., Scott, E.M., Pawson, P. and Nolan, A. 2001. Development of a behaviour-based scale to measure pain in dogs. Veterinary Record 148, 525-31.

Holton, L.L., Scott, E.M., Nolan, A.M., Reid, J., Welsh, E. and Flaherty, D. 1998a. Comparison of three methods of pain scoring used to assess clinical pain in dogs. Journal of the American Veterinary Medical Association 212, 61-6.

Holton, L.L., Scott, E.M., Nolan, A.M., Reid, J. and Welsh, E. 1998b. Relationship between physiological factors and clinical pain in dogs scored using a numerical rating scale. Journal of Small Animal Practice 39, 469-74.

Home Office. 1989. Animal (Scientific Procedures) Act 1986. Code of practice for the housing and care of animals used in scientific procedures. Her Majesty's Stationary Office, London, UK.

Hopkins, S.G., Schubert, T.A. and Hart, B.L. 1976. Castration of adult male dogs: Effects on roaming, aggression, urine marking and mounting. Journal of the American Veterinary Medical Association 168, 1108-10.

Houpt, K.A. 1991. Feeding and drinking behaviour problems. Veterinary Clinics of North America: Small Animal Practice, 21, 281-98.

Houpt, K.A. 1996. Ingestive behaviour: The control of feeding in dogs and cats. IN Readings in companion animal behaviour, edited by V.L. Voith and P.L. Borchelt. Veterinary Learning Systems, Trenton, USA. Pp. 198-206.

Howe, L.M., slater, M.R., Boothe, H.W., Hobson, H.P., Holcom, J.L. and Spann, A.C. 2001. Long-term outcomes of gonadectomy performed at an early age or traditional age in dogs. Journal of the American Veterinary Medical Association 218, 217-21.

Hsu, Y., Severinghaus, L.L. and Serpell, J.A. 2003. Dog keeping in Taiwan: Its contribution to the problem of free-roaming dogs. Journal of Applied Animal Welfare Science 6, 1-23.

Hubrecht, R.C. 1993. A comparison of social and environmental enrichment methods for laboratory housed dogs. Applied Animal Behaviour Science 37, 345-61.

Hubrecht, R. 1995a. The welfare of dogs in human care. IN The Domestic Dog, its evolution, behaviour and interaction with people, edited by J. Serpell. Cambridge University Press, Cambridge, England. Pp. 179-98.

Hubrecht, R.C. 1995b. Enrichment in puppyhood and its effects on later behaviour of dogs. Laboratory Animal Science 45, 70-5.

Hubrecht, R. 2002. Comfortable quarters for Dogs in Research Institutions. IN Comfortable Quarters for Laboratory Animals, edited by V. Reinhardt and A. Reinhardt. Animal Welfare Institute, Washington, USA. Pp. 56-64.

Hubrecht, R. 2004. The welfare of laboratory dogs. IN The Welfare of Laboratory Animals, edited by E. Kaliste. Springer Textbooks, Germany.

Hubrecht R.C., Serpell, J.A. and Poole, T.B. 1992. Correlates of pen size and housing conditions on the behaviour of kennelled dogs. Applied Animal Behaviour Science 34, 365-83.

Hudson, J.T., Slater, M.R., Taylor, L., Scott, H.M. and Kerwin, S.C. 2004. Assessing repeatability and validity of a visual analogue scale questionnaire for use in assessing pain and lameness in dogs. American Journal of Veterinary Research 65, 1634-43.

Hughes, J., Smith, T.W., Kosterlitz, H.W., Fothergill, L.A., Morgan, B.A. and Morris, H.A. (1975). Identification of two related pentapeptides from the brain with potent opiate agonist activity. Nature 258, 577-9.

Hunthausen, W. 1997. Effects of aggressive behaviour on canine welfare. Journal of the American Veterinary Medical Association 210, 1134-6.

Ilkiew, J.E., Davis, P.E. and Church, D.B. 1989. Hematologic, biochemical, blood-gas, and acid-base values in Greyhounds before and after exercise. American Journal of Veterinary Research 50, 583-6.

Inserro, J.C. 1997. Help prospective owners spot potential pet problems. Journal of the American Veterinary Medical Association 210, 470-1.

Institute of Laboratory Animal Resources. 1996. Guide for the Care and Use of Laboratory Animals. National Academy Press, Washington DC, USA.

Ireland, B.W. 1998. Race track biomechanics and design. IN Canine sports medicine and surgery, edited by M.S. Bloomberg, J.F. Dee, R.A. Taylor and J.R. Gannon. W.B. Saunders Company, Philadelphia, USA. Pp. 412-415.

Irvine, L. 2003. The problem of unwanted pets: A case study of how institutions 'think' about client's needs. Social Problems 50, 550-66.

Jagoe, A. and Serpell, J. 1996. Owner characteristics and interaction and the prevalence of canine behaviour problems. Applied Animal Behaviour Science 47, 31-42.

Jagoe, J.A. and Serpell, J.A. 1988. Optimum time for neutering. Veterinary Record 447.

Janssens, L.A.A. and Janssens, G.H.R.R. 1991. Bilateral flank ovariectomy in the dog – surgical technique and sequelae in 72 animals. Journal of Small Animal Practice 32, 249-52.

Jeffels, W. 1997. Frozen tail or limber tail in working dogs. Veterinary Record 140, 564.

Jennings, P.B., Moe, J.B., Elwell, P.A., Sands, L.D. and Stedman, M.A. 1971. Survey of diseases of military dogs in the Republic of Vietnam. Journal of the American Veterinary Medical Association 159, 434-40.

Jennings, P.B. and Freeman, J.R. 1998. Veterinary Problems unique to security and detector dogs. IN: Canine sports medicine and surgery, edited by M.S. Bloomberg, J.F. Dee, R.A. Taylor and J.R. Gannon. W.B. Saunders Company, Philadelphia, USA. Pp. 434-42.

Johnson, J.M. 1991. The veterinarian's responsibility" Assessing and managing acute pain in dogs and cats. Part I. Compendium of Continuing Education for the Practicing Veterinarian 13, 804-8.

Johnston, S.A. 1996. Physiology, mechanisms and identification of pain. IN Predictable Pain Management. Proceedings of a Symposium of the North American Veterinary Conference, 5-12.

Jones, S. 2003. Research Defense Society (RDS) Newsletter (winter), 11.

Joubert, K.E. 2001. The use of analgesic drugs by South African veterinarians. Journal of the South African Veterinary Association 72, 57-60.

Juarbe-Diaz, S.V. 1997. Social dynamics and behaviour problems in multiple-dog households. Veterinary Clinics of North America: Small animal Practice 27, 497-514.

Juarbe-Diaz, S.V. and Houpt, K.A. 1996. Comparison of two antibarking collars for treatment of nuisance barking. Journal of the American Animal Hospital Association 32, 231-5.

Kaninski, J., Call, J. and Fischer, J. 2004. Word learning in a domestic dog: evidence for "fast mapping". Science 304, 1682-3.

Katz, S. 1976. Pets and People. The politics of emotion without reason. IN Proceedings of the First Canadian symposium on pets and society, edited by F.M. Loew, A. McWilliam and H.C. Rowsell. Canadian Veterinary Medical Association, Ottawa, Canada.

Kealy, R.D., Olsson, S.E., Monti, K.L., Lawler, D.F., Biery, D.N., Helms, R.W., Lust, G. and Smith, G.F. 1992. Effects of limiting food consumption on the incidence of hip dysplasia in growing dogs. Journal of the American Veterinary Medical Association 201, 857-63.

Kealy, R.D., Lawler, D.F., Ballam, J.M., Mantz, S.L., Biery, D.N., Greeley, E.H., Lust, G., Segre, M., Smith, G.K. and Stowe, H.D. 2002. Effects of diet restriction of life span and age-related changes in dogs. Journal of the American Veterinary Association 220, 1315-20.

Keep, J.M. 1970. Gunshot injuries in dogs and cats. Australian Veterinary Journal 46, 330-4.

Keinzle, E., Bergler, R. and Mandernach, A. 1998. A comparison of the feeding behaviour and the human-animal relationship in owners of normal and obese dogs. The Journal of Nutrition 128, 2779S-82S.

Ketteritzsch, K., Haman, H., Brahm, R., Grubendorf, H., Rosenhagen, C.U. and Dist, O. 2004. Genetic analysis of presumed inherited eye diseases in Tibetan terriers. The Veterinary Journal 168, 151-9.

Kharmachi, H., Haddad, N. and Matter, H. 1992. Tests of four baits for oral vaccination of dogs against rabies in Tunisia. Veterinary Record 130, 494.

Kidd, A.H. and Kidd, R.M. 1989. Factors in adults' attitudes towards pets. Psychological Reports 65, 905-40.

Kidd, A.H. and Kidd, R.M. 1990. Factors in children's attitudes towards pets. Psychological Reports 66, 775-86.

Kidd, A.H., Kidd, R.M. and George, C.G. 1992a. Successful and unsuccessful pet adoptions. Psychological Reports 70, 547-61.

Kidd, A.H., Kidd, R.M. and George, C.G. 1992b. Veterinarians and successful pet adoptions. Psychological Reports 71, 551-7.

Kienle, R.D., Thomas, W.P. and Pion, P.D. 1994. The natural clinical history of canine congenital subaortic stenosis. Journal of Veterinary Internal Medicine 8, 423-31.

Kiler-Matznick, J., Brisbin, I.L., Feinstein, M. and Bulmer, S. 2003. An updated description of the New Guinea singing dog (*Canis hallstromi*, Trooughton, 1957). Journal of Zoology, London 261, 109-18.

Kirk, R.W. 1986. A catalogue of congenital and hereditary disorders of dogs (by breeds). IN Current veterinary Therapy IX, Small Animal Practice, edited by R.W. Kirk. W.B. Saunders Co., Philadelphia, USA. Pp. 1281-5.

Kitala, P.M., Mcdermott, J.J., Kyule, M.N. and Cathuma, J.M. 1993. Features of dog ecology relevant to rabies spread in Machakos District, Kenya. Onderstepoort Journal of Veterinary Research 60, 445-9.

Klassen, B., Buckley, J.R. and Esmail, A. 1996. Does the dangerous dog act protect against animal attacks: A prospective study of mammalian bites in the accident and emergency department. Injury-International Journal of the care for the injured 27, 89-91.

Klinckmann, G., Koniszewski, G. and Wegner, W. 1986. Light microscopic investigations on the retina of dogs carrying the merle factor. Journal of Veterinary Medicine Series A 33, 674-88.

Knight, L.J. 1984. A guide to training sheep dogs in New Zealand. Gainsborough Printing Co, Auckland, New Zealand.

Kobelt, A.J. 2004. The behaviour and welfare of pet dogs in suburban backyards. Unpublished PhD thesis. University of Melbourne, Australia.

Kobelt, A.J., Hempsworth, P.H., Barnett, J.L. and Butler, K.L. 2003a. Source of sample variation in saliva cortisol in dogs. Research in Veterinary Science 75, 157-61.

Kobelt, A.J., Hemsworth, P.H., Barnett, J.L. and Coleman, G.J. 2003b. A survey of dog ownership in suburban Australia – conditions and behaviour problems. Journal of Applied Animal Behaviour Science 82, 137-48.

Koerner, H. 1998. Raising, training and conditioning the racing greyhound. IN Canine sports medicine and surgery, edited by M.S. Bloomberg, J.F. Dee, R.A. Taylor and J.R. Gannon. W.B. Saunders Company, Philadelphia, USA. Pp. 363-5.

Kogan, L.R. and Viney, W. 1998. Reported strength of human animal bonding and methods of acquiring a dog. Psychological Reports 82, 647-50.

Kogure, N. and Yamazaki, K. 1990. Attitudes to animal euthanasia in Japan: a brief review of cultural influences. Anthrozoos 3, 151-4.

Kohn, D.F. 1995. Dogs. IN The Experimental Animal in Biomedical Research Volume 11, edited by B.E. Rollin and M.L. Kesel. CRC Press, London, UK. Pp. 435-56.

Koler-Matznick, J., Lehhr Brisbin, J.I., Feinstein, M. and Bulmer, S. 2003. An updated description of the New Guinea Dog (*Canis hallstromi*, Troughton 1957). Journal of Zoology (London) 261, 109-118.

Kornas, S, Nowosad, B. and Skalska, M. 2002. Hookworm infection in dogs in stray animal shelters. Medycyna Weterynaryjna 58, 291-4.

Kruuk, H. and Snell, H. 1981. Prey selection by feral dogs from a population of marine iguanas (*Amblyrhynchus cristatus*). Journal of Applied Ethology 18, 197-204.

Kuehn, B.M. 2002. Animal Hoarding. Journal of the American Veterinary Medical Association 221, 1087-9.

Kuwert et al. 1985. Rabies in the tropics ist editiono edited by wnadeler id.

Kyles, A.E., Hardie, E.M., Hansen, B.D. and Papich, M.G. 1998. Comparison of transdermal fentanyl and intramuscular oxymorphone on post-operative pain behaviour after ovariohysterectomy in dogs. Research in Veterinary Science 65, 245-51.

Kyono, M. 2002. The New Zealand Police Dogs. Unpublished MSc thesis, Massey University, New Zealand.

Land, T.W. 2000. Favors early spay/neuter. Journal of American Veterinary Medical Association 216, 659-60.

Landsberg, G., Hunthausen, W. and Ackerman, L. 1997. Handbook of behaviour problems of the dog and cat. Butterworth Heinemann, Oxford, England.

Lane, D.R., McNicholas, J. and Collis, G.M. 1998. Dogs for the disabled: benefits to recipients and welfare of the dog. Applied Animal Behaviour Science 59, 49-60.

Langley, J. 1992. The incidence of dog bites in New Zealand. New Zealand Medical Journal 105, 33-5.

Laredo, F.G., Belda, E., Murciano, J., Escobar, M., Navarro, A., Robinson, K.J. and Jones, R.S. 2004. Comparison of the analgesic effects of meloxicam and carprofen administered preoperatively to dogs undergoing orthopaedic surgery. The Veterinary Record 155, 667-71.

Lascalles, B.D.X., Butterworth, S.J. and Waterman, A.E. 1994. Post-operative analgesic and sedative effects of carprofen and pethidine in dogs. Veterinary Record 134, 187-91.

Lascelles, B.D.X., Cripps, P.J., Jones, A. and Waterman, A.E. 1997. Post-operative hypersensitivity and pain: the pre-emptive value of pethidine for ovariohysterectomy. Pain 73, 461-71.

Lascelles, B.D.X. and Main, D.J.C. 2002. Surgical trauma and chronically painful conditions–within our comfort level but beyond theirs? Journal of the American Veterinary Medical Association 221, 215-22.

Le Nobel, W.E., Robben, S.R.M, Dopfer, D., Hendrikx, W.M.L., Boersema, J.H. and Eysker, F.F.M. 2004. Infections with endoparasites in dogs in Dutch aniamal shelters. Tijdschrift voor Diergeneeskunde 129, 40-4.

Leadon, D.P. and Mullins, E. 1991. Relationship between kennel size and stress in greyhounds transported short distances by air. Veterinary Record 129, 70-3.

Ledger, R. 1997. Understanding owner-dog compatibility. International Veterinarian 9, 17-23.

Ledger, R.A. 1999. Owner and dog characteristics: their effects on the success of the owner-dog relationship. Part 1: owner attachment and ownership success. Veterinary International 11, 2-10.

Lee, J.A., Hinchcliff, K.W., Piercy, R.J., Schmidt, K.E. and Nelson, S. 2004. Effects of racing and nontraining on plasma thyroid hormone concentrations. Journal of the American Veterinary Medical Association 224, 226-31.

Lee, R.B. 1979. The !Kung San: Men, women and work in a foraging society. Cambridge University Press, Cambridge, England.

Leighton, E.A. 1997. Genetics of canine hip dysplasia. Journal of the American Veterinary Medical Association 210, 1474-9.

Leney, J. and Remfry, J. 2000. Dog population management. IN Dogs, Zoonoses and Public Health, edited by C.N.L. Macpherson, F.X. Meslin and A.I. Wandeler. CABI Publishing, Wallingford, UK. Pp. 299-332.

Leonard, J.A., Wayne, R.K., Wheeler, J., Valadez, R., Guillen, S. and Vila, C. 2002. Ancient DNA evidence for Old World origin of New World dogs. Science 298, 1613-6.

Leppanen, M. and Saloniemi, H. 1999. Controlling canine hip dysplasia in Finland. Preventive Veterinary Medicine 42, 121-31.

Lepper, M., Kass, P.H. and hart, L.A. 2002. Predictionof adoption versus euthanasia among dogs and cats in a California animal shelter. Journal of Applied Animal Welfare 5, 29-42.

Leslie, B.E., Meek, A.H., Kawash, G.F. and McKeown, D.B 1994. An epidemiological survey of pet ownership in Ontario. Canadian Veterinary Journal 35, 218-22.

Lester, S.J., Mellor, D.J., Holmes, R.J., Ward, R.N. and Stafford, K.J. 1996. Behavioural and cortisol responses of lambs to castration and tailing using different methods. New Zealand Veterinary Journal 44, 45-54.

Lever, C. 1985. Domestic Dog. IN Naturalised Mammals of the World. Longman, London, England. Pp. 49-54.

Ley, S.J., Livingston, A. and Waterman, A.E. 1989. The effect of chronic clinical pain on thermal and mechanical thresholds in sheep. Pain 39, 353-7.

Ley, S.J., Waterman A.E. and Livingston, A. 1995. A field study of the effect of lameness on mechanical nociceptive thresholds in sheep. The Veterinary Record 137, 85-7.

Lieberman, L.L. 1988. The optimum time for neutering surgery of dogs and cats. Veterinary Record 122, 369.

Linde-Forsberg, C. 2001. Reproduction and modern reproductive technology. IN The Genetics of the Dog, edited by A. Ruvinsky and J. Sampson. CABI Publishing, Wallingford, England. Pp. 461-485.

Lindhart, S.B., Baer, G.M., Torres, J.M.B., Engeman, R.M., Collins, E.F., Meslin, F.X., Schumacher, C.L., Taweel, A.H. and Wlodkowski, J.C. 1997. Acceptance of candidate baits by domestic dogs for delivery of oral rabies vaccine. Onderstepoort Journal of Veterinary Research 64, 115-24.

Lindsay, S.R. 2000. Handbook of applied dog behaviour and training. Volume One: Adaption and learning. Iowa State University Press, Iowa, USA.

Lindsay, S.R. 2001. Handbook of applied dog behaviour and training. Volume Two: Etiology and assessment of behaviour problems. Iowa State University Press, Iowa, USA.

Line, S.W., Hart, B.L. and Sanders, L. 1985. Effect of prepubertal versus postpubertal castration on sexual and aggressive behaviour in male horses. Journal of the American Veterinary Medical Association 186, 249-51.

Lipscomb, V.J., Lawes, T.J., Goodship, A.E. and Muir, P. 2001. Asymmetric densitometric and mechanical adaptation of the left fifth metacarpal bone in racing greyhounds. Veterinary Record 148, 308-11.

Lloyd, J.K.F. 2004. Exploring the match between people and their guide dogs. PhD Thesis, unpublished, Massey University, New Zealand.

Logan, E.I., Wiggs, R.B., Zetner, L. and Hefferren, J.J. Dental Disease. IN Small Animal Clinical Nutrition, edited by M.S. Hand, C.D. Thatcher, R.L. Remillard and P. Roudebush. Mark Morris Institute, Topeka, USA. Pp. 475-504.

Lohachit, C. and Tanticharoenyos, K. 1991. Variety methods used for controlling dog population. Part 1. Canine contraception in the male. The Journal of Thai Veterinary Practitioners 3, 119-128.

Lohachit, C. and Tanticharoenyos, K. 1992. Variety methods use for controlling dog population. Part 11. Canine contraception in female. The Journal of Thai Veterinary Practitioners 4, 57-71.

Lohachit, C. and Tanticharoenyos, K. 1997. Technique for vasectomy in dogs using human no scalpel vasectomy technique clamp. The Thai Journal of Veterinary Medicine 22, 147-156.

Lonsdale, T. 2001. Raw Meaty Bones. Rivetco P/L, New South Wales, Australia.

Lord, L.K., Wittum, T.E., Neer, C.A. and J.C. Gordon. 1998. Demographic and needs assessment survey of animal care and control agencies. Journal of the American Veterinary Medical Association 213, 483-7.

Love, M. and Overall, K.L. 2001. How anticipating relationships between dogs and children can help prevent disasters, Journal of the American Veterinary Medical Association 219, 446-51.

Loveridge, G. 1994. Provision of environmentally enriched housing for dogs. Animal Technology 45, 1-19.

Loveridge, G.G. 1998. Environmentally enriched dog housing. Applied Animal Behaviour Science 59, 101-13.

Luescher, A.U. 2003. Diagnosis and management of compulsive disorders in dogs and cats. The Veterinary Clinics: Small Animal Practice 33, 253-67.

Luescher, U.A. 1993. Animal Behaviour Case of the Month. Journal of the American Veterinary Medical Association 203, 1538-9.

Luescher, U.A., McKeown, D.B. and Halip, J. 1991. Stereotypic or obsessive-compulsive disorders in dogs and cats. Veterinary Clinics of North America: Small Animal Practice 21, 401-13.

Lynch, J.J. and Gantt, W. 1968. The heart rate component of the social reflex in dogs: the conditional effect of petting and person. Conditioned Reflex 3, 69-80.

Maarschalkerweerd, R.J., Endenburg, N., Kirpensteijn, J. and Knol, B.W. 1997. Influence of orchidectomy on canine behaviour. Veterinary Record 140, 617-9.

MacArthur, J.A. 1987. The dog. IN The UFAW handbook on the care and management of Laboratory Animals 6th edition, edited by T. Poole. Longman Scientific and Technical, London, UK. Pp. 456-75.

Macdonald, D.W. and Carr, G.M. 1995. Variation in dog society: between resource dispersion and social flux. IN The Domestic Dog; its evolution, behaviour and interactions with people, edited by J. Serpell. Cambridge University Press, Cambridge, England. Pp. 199-216.

McIlroy, J.C., Cooper, R.J., Gifford, E.J., Green, B.F. and Newgrain, K.W. 1986. The effect on wild dogs Canis familiaris of 1080 poisoning campaigns in Kosciusko National parks, NSW. Australian Wildlife Research 13, 534-44.

MacKay, C.A. 1993. Veterinary practitioners' role in pet overpopulation. Journal of the American Veterinary Medical Association 202, 918-21.

Mackenzie, S.A., Oltenacu, E.A.B. and Leighton, E. 1985. Heritability estimate for temperament scores in German Shepherd dogs and its genetic correlation with hip dysplasia. Behavior Genetics 15, 475-82.

Macpherson, C.N.L., Meslin, F.X. and Wandeler, A.I. 2000. Preface IN Dogs, Zoonoses and Public Health, edited by C.N.L. Macpherson, F.X. Meslin and A.I. Wandeler. CABI Publishing, Wallingford, UK. Pp. XI-XII.

Machado, L.E.P., Hurnik, J.F. and Burton, J.H. 1997. The effect of amniotic fluid ingestion on the nociception of cows. Physiology and Behaviour 62, 1339-44.

Mahlow, J.C. 1999. Estimation of the proportions of dogs and cats that are surgically desexed. Journal of the American Veterinary Medical Association 215, 640-3.

Malaga, H., Garcia, A., Urdaneta, N., Gomezbarrios, F. and Bocaranda, G. 1992. Can rabies be eradicated – the epidemiological basis for urban control in Venezuela. Health Policy and Planning 7, 279-83.

Markwell, P.J., Butterwick, R.F., Wills, J.M. and Raiha, M. 1994. Clinical-studies in the management of obesity in dogs and cats. International Journal of Obesity 18, S39-S43.

Markwell, P.J., Erk, W., Parkin, G.D., Sloth, C.J. and Kristensen, T.S. 1990. Obesity in the dog. Journal of Small Animal Practice 31, 533-7.

Marschark, E.D. and Baenninger, R. 2002. Modification of instinctive herding dog behaviour using reinforcement and punishment. Anthrozoos 15, 51-68.

Marsh, F.O. 1994. Global awareness trends affecting petfood manufacturers. Petfood Industry 36, 4-19.

Marston, L.C., Bennett, P.C. and Coleman, G.J. 2004. What happens to shelter dogs? An analysis of data for 1 year from three Australian shelters. Journal of Applied Animal Welfare 7, 27-47.

Marx, M.B. and Furculow, M.L. 1969. What is the dog population? A review of surveys in the United States. Archives of Environmental Health 19, 217-9.

Mason, B.J.E. 1991. Control of fighting dogs. Veterinary Record 128, 553.

Mason, E. 1970. Obesity in pet dogs. Veterinary Record 86, 612-6.

Mason, G.J. 1991. Stereotypes and suffering. Behavioural Processes 25, 103-15.

Mathews, K.A., Paley, D.M., Foster, R.A., Valliant, A.E.and Young, S.S. 1996. A comparison of ketorolac with flunixin, butorphanol and oxymorphone in controlling postoperative pain in dogs. Canadian Veterinary Journal 37, 557-67.

Matter, H.C. and Daniels, T.J. 2000. Dog ecology and population biology. IN Dogs, Zoonoses and Public Health, edited by C.N.L. Macpherson, F.X. Meslin and A.I. Wandeler. CABI Publishing, Wallingford, UK. Pp. 17-62.

Matter, H.C., Wandeler, A.L., Neuenschwander, B.E., Harischandra, L.P.A. and Meslin, F.X. 2000. Study of the dog population and the rabies control activities in the Mirigama area of Sri Lanka. Acta Tropical 75, 95-108.

McCobb, E.C., Brown, E.A., Damiani, K. and Dodman, N.H. 2001. Thunderstorm phobia in dogs: An internet survey of 69 cases. Journal of the American Animal Hospital Association 37, 319-24.

McCormick, R.H. 1999. Comparing ear cropping, tail docking and neutering. Journal of the American Veterinary Medical Association 215, 926.

McCrave, E.A. 1991. Diagnostic criteria for separation anxiety in the dog. The Veterinary Clinics of North America; Small animal practice 21, 247-55.

McCreath, C.P.F. 1993. Docking of dogs. Veterinary Record 133, 303.

McGreevy, P.D. and Nicholas, F.W. 1999. Some practical solutions to welfare problems in dog breeding. Animal Welfare 8, 329-41.

McLean, I.G. 2001. Designer dogs: Improving the quality of mine detection dogs. Geneva International Centre for Humanitarian Demining, Geneva, Switzerland.

McMillan, F.D. 2000. Quality of life in animals. Journal of the American Veterinary Medical Association 216, 1904-10.

McNamara, J.H. 1972. Nutrition for military working dogs under stress. Veterinary Medicine and Small Animal Clinician 67, 615-23.

Meenken, D. and Booth, L.H. 1997. The risk to dogs of poisoning from socium monofluoroacetate (1080) residues in possum (Trichosurus vulpecula). New Zealand Journal of Agricultural Research 40, 573-6.

Meggitt, M.J. 1965. The association between Australian Aborigines and Dingoes. IN Man, culture and animals, edited by A. Leeds and A.P. Vayda. American Association for the Advancement of Science, 78, 7-27.

Mellor, D.J. and Stafford, K.J. 1999. Assessing and minimising the distress caused by painful husbandry procedures. In Practice 21, 436-46.

Mellor, D.J. and Stafford, K.J. 20000). Acute castration and/or tailing distress and its alleviation in lambs. New Zealand Veterinary Journal 48, 33-43.

Mellor, D.J., Cook, C.J. and Stafford, K.J. 2000. Quantifying some responses to pain as a stressor. The Biology of Animal Stress, edited by G.P. Moberg and J.A. Mench. CABI Publishing, Wallingford, UK. Pp. 171-98.

Mellor, D.J., Stafford, K.J., Todd, S.E., Lowe, T.E., Gregory, N.G., Bruce, R.A. and Ward, R.N. 2002. A comparison of catecholamine and cortisol responses of young lambs and calves to painful husbandry procedures. Australian Veterinary Journal 80, 228-33.

Meltzer, M.L. and Rupprecht, C.E. 1998. A review of the economics and control of rabies part 2: Rabies in dogs, livestock and wildlife. Pharmacoeconomics 14, 481-98.

Mercer, P. 1992. Docking of dogs. Veterinary Record 131, 375.

Merskey, H. (1979). Pain terms: a list with definitions and notes on usage. Pain 6, 249-52.

Mertens, P.A. and Unshelm, J. 1996. Effects of group and individualhousing on the behaviour of kennelled dogs in animal shelters. Anthrozoos 9, 40-51.

Meslin, F.X., Miles, M.A., Vexenat, J.A. and Gemmell, M.A. 2000. Zoonoses control in dogs. IN Dogs, Zoonoses and Public Health, edited by C.N.L. Macpherson, F.X. Meslin and A.I. Wandeler. CABI Publishing, Wallingford, UK. Pp. 333-72.

Miklosi, A., Polgardi, R., Topal, J. and Csanyi, V. 1998. Use of experimenter-given cues in dogs. Animal Cognition 1, 113-21.

Miklosi, A.

Miklosi, A., Kubinti, E., Topal, J., Gacsi, M. Viranyi, Z. and Csanyi, V. 2003. A simple reason for a big difference: Wolves do not look back at humans, but dogs do. Current Biology, 13, 763-6.

Miller, D.D., Staats, S.R. Partlo, C. and Rada, K. 1996. Factors associated with the decision to surrender a pet to an animal shelter. Journal of the American Veterinary Medical Association 209, 738-42.

Miller, G.Y. 1998. Earnings, feminisation and consequences for the future of the veterinary profession. Journal of the American Veterinary Medical Association 213, 340-4.

Millot, J.L., Filiatre, J.C., Gagnon, A.C., Eckerlin, A.and Montagner, H. 1988. Children and their pet dogs: how they communicate. Behavioural Processes 17, 1-15.

Millot, J.L. 1994. Olfactory and visual cues in the interaction systems between dogs and children. Behavioural Processes 33, 177-88.

Mills, D.S. Gandia Estelles, M., Coleshaw, P.H. and Shorthouse, C. 2003. Retrospective analysis of the treatment of foreworks fear in dogs. Veterinary Record 153, 561-2.

Molony, V. and Kent, J.E. 1997. Assessment of acute pain in farm animals using behavioural and physiological measurements. Journal of Animal Science 75,138-42.

Molony, V., Kent, J.W. and McKendrick, I.J. 2002. Validation of a method for assessment of an acute pain in lambs. Applied Animal Behaviour Science 76, 215-38.

Molony, V., Kent, J.E., Hosie, B.D. and Graham, M.J. 1997. Reduction in pain suffered by lambs at castration. The Veterinary Journal 153, 205-13.

Moore, G.E., Burkman, K.D., Carter, M.N. and Peterson, M.R. 2001. Causes of death or reasons for euthanasia in military working dogs: 927 cases. Journal of the American Veterinary Medical Association 219, 209-14.

Moritz, A., Walcheck, B.K., Deye, J. and Weiss, D.J. 2003. Effects of short-term racing activity on platelet and neutrophil activation in dogs. American Journal of Veterinary Research 64, 855-9.

Morton, D. 1992, Docking of dogs: practical and ethical issues. Veterinary Record 131, 301-6.

Morton, D.B. and Griffiths, P.H.M. (1985). Guidelines on the recognition of pain, distress and discomfort in experimental animals and hypothesis for assessment. Veterinary Record 116, 431-6.

Mozdy, A.D. 1997. Pay attention Rover. The New Scientist XX, 30-33.

Mugford, R.A. 1987. The influence of nutrition on canine behaviour. Journal of Small Animal Practice 28, 1046-55.

Mugford, R.A. 1991. Where to put your choker. International Journal for the study of Animal Problems 2, 249-51.

Mugford, R.A. 1995. Canine behavioural therapy. IN The domestic dog its evolution, behaviour and interaction with people, edited by J. Serpell. Cambridge University Press, Cambridge, England. Pp. 139-52.

Muir, W.W. and Woolf, C.J. 2001. Mechanisms of pain and their therapeutic implications. Journal of the American Veterinary Medical Association 219, 1346-56.

Muir, W.W., Wiese, A.J. and Wittum, T.E. 2004. Prevalence and characteristic of pain in dogs and cats examined as outpatients at a veterinary teaching hospital. Journal of the American Veterinary Medical Association 224, 1459-63.

Munro, H.M.C. and Thrusfield, M.V. 2001a. 'Battered Pets': features that raise suspicion of non-accidental injury. Journal of Small Animal Practice 42, 218-26.

Munro, H.M.C. and Thrusfield, M.V. 2001b. 'Battered Pets' injuries found in dogs and cats. Journal of Small Animal Practice 42, 279-90.

Munro, H.M.C. and Thrusfield, M.V. 2001c. 'Battered Pets': sexual abuse. Journal of Small Animal Practice 42, 333-7.

Munro, H.M.C. and Thrusfield, M.V. 2001d. 'Battered Pets': Munchausen syndrome by proxy (factitious illness by proxy). Journal of Small Animal Practice 42, 385-9.

Murphy, L.A., Gwaltney-Brant, S.M., Albretsen, J.C. and Wismer, T.A. 2003. Toxicological agents of concern for search-and-rescue dogs responding to urban disasters. Journal of the American Veterinary Medical Association 222, 296-304.

Murray, R.W. 1992. Veterinarians and urban animal management. Australian Veterinary Practitioner 22, 134-8.

Murray, R.W. 1993. Urban Animal Problems. IN Animal Behaviour. Proceedings 214 of the Post-Graduate Committee in Veterinary Science, University of Sydney. Pp. 1-13.

Murray, R.W. and Speare, R. 1995. Unwanted pets: Disposal of dogs and cats in a provincial Australian city. Australian Veterinary Practitioner 25, 69-73.

Musil, R. 1984. The first known domestication of wolves in central Europe. IN Animals and Archaeology: 4. Husbandry in Europe, edited by C. Grigson and J. Clutton-Brock. BAR International, Oxford, UK, Series 227, 23-6.

Myles, S. 1991. Trainers and chokers; how dog trainers affect behavioural problems in dogs. Veterinary Clinics of North America: Small Animal Practice 21, 239-46.

Naderi, S., Miklosi, A., Doka, A. and Csani, V. 2001. Co-operative interactions between blind persons and their dogs. Applied Animal Behaviour Science 74, 59-80.

Nash, A.S. and Watson, I. 1994. Smoke detectors: effect of low battery warning soundon dogs. Veterinaary Record January 22, 100.

Nasser, R., Mosier, J.E. and Williams. L.W. 1984. Study of the canine and feline populations in the greater Las Vegas area. American Journal of Veterinary Research 45, 282-7.

National Health and Medical Research Council. 1996. Policy on the care of dogs in medical research. NHMRC, Canberra, Australia.

Neal, K. 1992. Docking of dogs. Veterinary Record 131, 399.

Neamand, J., Sweeny, W.T., Creamer, A.A. and Conti, P.A. 1975. Cage activity in the laboratory Beagle; a preliminary study to evaluate a method of comparing cage size to physical activity. Laboratory Animal Science 25, 180-3.

Neilson, J.C., Eckstein, R.A. and Hart, B.L. 1997. Effects of castration on problem behaviours in male dogs with reference to age and duration of behaviour. Journal of the American Veterinary Medical Association 211, 180-2.

Nemcova, D. and Novak, P. 2003. Adoption of dogs in the Czech Republic. Acta Veterinarian Brno 72, 421-7.

Netto, W.J. and Planta, D.J.U. 1997. Behavioural testing for aggression in the domestic dog. Applied Animal Behaviour Science 52, 243-63.

Netto, W.J., van der Borg, J.A.M. and Planta, D.J.U. 1990. Behaviour tests for dogs at animal shelters. Tijdschrift voor Diergeneeskunde 115, 7-8S.

Neville, P. 1997. Canine anorexia – an interesting case. The Veterinary Nursing Journal 12, 20.

New, J.C. Jr., Salman, M.D., Scarlett, J.M., Kass, P.H., Vaughn, J.A., Scherr, S. and Kelch, W.J. 1999. Moving: Characteristics of dogs and cats and those relinquishing them to 12 U.S. animal shelters. Journal of Applied Animal Welfare Science 2, 83-96.

Newby, J. 1997. The pact for survival: humans and their animal companions. ABC Books, Sydney, Australia.

XXXXNHMRC, CSIRO, ACC (1990). Australian code of practice for the care and use of animals for scientific purposes. Australian Government Publishing Services, Canberra.

Nicholas, F.W. and Thomson, P.C. 2004. Inherited disorders: sustained attack from several quarters. The Veterinary Journal 168, 114-5.

Nicholson, J., Kemp-Weeler, S. and Griffiths, D. 1995 Distress arising from the end of a guide dog partnership. Anthrozoos 8, 100-10.

Nolan, A. (2000a). Patterns and management of pain in animals. IN Pain: its nature and management in man and animals, edited by Lord Soulsby of Swaffham and D Morton. The Royal Society of Medicine Press, London, UK. Pp. 93-9.

Nolan, A. and Reid, J. 1993. Comparison of the postoperative analgesic and sedative effects of carprofen and papaveretum in the dog. Veterinary Record 133, 240-2.

Nolan, A.M. 2000. Pharmacology of analgesic drugs. IN Pain Management in Animals, edited by P. Flecknell and A. Waterman-Pearson. WB Saunders. London, UK.

Nolan, R.S. AVMA adopts position on ear cropping and tail docking – Veterinarians should counsel owners of risk before surgery. Journal of the American Veterinary Association 215, 461-2.

Nold, J.L., Peterson, L.J. and Fedde, M.R. 1991. Physiological changes in the running greyhound (Canis familiaris) : influence of race length. Comparative Biochemistry and Physiology 100A, 623-7.

Noonan, G.J., Rand, J.S., Blackshaw, J.K. and Priest, J. 1996. Behavioural observations of puppies undergoing tail docking. Applied Animal Behaviour Science 49, 335-42.

Noriega, P.B. and Lin, L.C. 2002. The growth of the gaming industry: The survival of greyhound racing in the State of Florida. Journal of Travel and Tourism 12, 39-57.

Norris, M.P. and Beaver, B.V. 1993. Application of behaviour therapy techniques to the treatment of obesity in companion animals. Journal of the American Veterinary Medical Association 202, 728-30.

Novinger, M.S., Sullivan, P.S. and McDonald, T.P. 1996. Determination of the lifespan of erythrocytes from greyhounds, using an in vitro biotinylation technique. American Journal of Veterinary Research 57, 739-42.

O'Connor, T.P. 1997. Working at relationships: Another look at animal domestication. Antiquity 71, 149-56.

O'Farrell, V. 1992. Manual of Canine Behaviour. British Small Animal Veterinary Association, England.

O'Farrell, V. 1995. Effects of owner personality and attitude on dog behaviour. IN The Domestic Dog , its evolution, behaviour and interactions with people, edited by J. Serpell. Cambridge University Press, Cambridge, England. Pp. 153-60.

O'Farrell, V. 1997. Owner attitudes and dog behaviour problems. Applied Animal Behaviour Science 52, 205-13.

O'Farrell, V. and Peachey, E. 1990. Behavioural effects of ovariohysterectomy on bitches. Journal of Small Animal Practice 31, 595-8.

Oberbauer, A.M. and Sampson, J. 2001. Pedigree analysis and genetic counselling. IN The Genetics of the Dog, edited by A. Ruvinsky and J. Sampson. CABI Publishing, Wallingford, England. Pp. 461-85.

Odendaal, J.S.J. 1994. Demographics of companion animals in South Africa. Journal of the South African Veterinary Association 65, 67-72.

Odendaal, J.S.J. and Meintjes, R.A. 2003. Neurophysiological correlates of affiliative behaviour between humans and dogs. The Veterinary Journal 165, 296-301.

Ogburn, P., Crouse, S., Martin, F. and Houpt, K.A. 1998. Comparison of behavioural and physiological responses of dogs wearing two different types of collars. Applied Animal Behavioural Science 61, 133-42.

Okkens, A.C., Kooistra, H.S. and Nickel, R.F. 1997. Comparison of long-term effects of ovariectomy versus ovariohysterectomy in bitches. Journal of Reproduction and Fertility 51, 227-31.

Olson, P.N. and Johnston, S.D. 1993. New developments in small animal population control. Journal of the American Veterinary Medical Association 202, 904-9.

Olsen, S.J. 1985. Origins of the domestic dog: The fossil record. The University of Arizona Press, Tuscon , USA.

Ott, R.S. 1996. Animal selection and breeding techniques that create diseased populations and compromise welfare. Journal of the American Veterinary Medical Association, 208, 1969-96.

Ottewill, D. 1968. Planning and design of accommodation for experimental dogs and cats. Laboratory Animal Symposia 1, 97-112.

Overall, K.L. 1997. Clinical behavioral medicine for small animals. Mosby, St. Louis, Missouri, USA.

Overall, K.L., Dunham, A.E. and Frank, D. 2001. Frequency of non-specific clinical signs in dogs with separation anxiety, thunderstorm phobia, and noise phobia, alone or in combination. Journal of the American Veterinary Medical Association 219, 467-73.

Overgaauw, P.A.M. and van Knapen, F. 2000. Dogs and nematode zoonoses. IN Dogs, Zoonoses and Public Health, edited by C.N.L. Macpherson, F.X. Meslin and A.I. Wandeler. CABI Publishing, Wallingford, UK. Pp. 213-56.

Padgett, G.A. 1998. Control of canine genetic diseases. Howell Book House, New York, USA.

Padgett, G.A., Madewell, B.R., Keller, E.T., Jodar, L. and Packard, M. 1995. Inheritance of histiocytosis in Bernese Mountain Dogs. Journal of Small Animal Practice 36, 93-8.

Pain, S. 1997. Plague dogs. The New Scientist 154, 32-37.

Pal, S.K. 2001. Population ecology of free-ranging dogs in West Bengal, India. Acta Theriologica 46, 69-78.

Pal, S.K. 2003a. Reproductive behaviour of free-ranging rural dogs in West Bengal, India. Acta Theriologica 48, 271-81.

Pal, S.K. 2003b. Population ecology of free-ranging dogs in West Bengal, India. Acta Theriologica 46, 67-78.

Pal, S.K. 2003c. Urine marking by free-ranging dogs (Canis familiaris) in relation to sex, season, place and posture. Applied Animal Behaviour Science 80, 45-9.

Pal, S.K., Ghosh and Roy, S. 1998a. Dispersal behaviour of free-ranging dogs (Canis familiaris) in relation to age, sex and dispersal distance. Applied Animal Behaviour Science 61, 123-32.

Pal, S.K., Ghosh and Roy, S. 1998b. Agonistic behaviour of free-ranging dogs (Canis familiaris) in relation to season, sex and age. Applied Animal Behaviour Science 61, 331-48.

Palatnik-de-Sousa, C.B., dos Santos, W.R., Franca-Silva, J.C., da Costa, R.T., Reis, A.B., Palatnik, N., Mayrink, W. and Genaro, O. 2001. Impact of canine control on the epidemiology of canine and human visceral leishmaniasis in Brazil. American Journal of Tropical Medicine and Hygeine 65, 5120-7.

Panciera, D.L., Hinchcliff, K.W., Olson, J. and Constable, P.D. 2003. Plasma thyroid hormone concentrations in dogs competing in a long-distance sled dog race. Journal of Veterinary Internal Medicine 17, 593-6.

Panichabhongee, P. 2001. The epidemiology of Rabies in Thailand. MSc thesis unpublished Massey University, New Zealand.

Parker, H.G., Kim, L.V., Sutter, N.B., Carlson, S., Lorentzen, T.D., Malek, T.B., Johnson, G.S., DeFrance, H.B., Ostrander, E.A. and Kruglyak, L. 2004. Genetic structure of the purebred domestic dog. Science 304, 1160-4.

Pascoe, P.J. 2002. Akternative methods of pain control. Journal of the American Veterinary Medical Association 221, 222-9.

Patronek, G.J. 1999. Hoarding of animals: An under-recognised public health problem in a difficult-to-study population. Public Health Reports 114, 81-7.

Patronek, G.J. and Rowan, A.N. 1995. Determining dog and cat numbers and population dynamics. Anthrozoos 8, 199-205.

Patronek, G.J., Beck, A.M. and Glickman, L.T. 1997. Dynamics of dog and cat populations in a community. Journal of the American Veterinary Medical Association 210, 637-42.

Patronek, G.J., Glickman, L.T. and Moyer, M.R. 1995. Population dynamics and the risk of euthanasia for dogs in an animal shelter. Anthrozoos 8, 31-43.

Patronek, G.J., Lawrence, T., Glickman, L.T., Beck, A.M., McCabe, G.P. and Ecker, C. 1996. Risk factors for relinquishment of dogs to an animal shelter. Journal of the American Veterinary Medical Association 209, 572-81.

Patterson, D.F. 2000. Companion animal medicine in the age of medical genetics. Journal of Veterinary Internal Medicine 14, 1-9.

Paul, E.S. and Podberscek, A.L. 2000. Veterinary education and students' attitudes towards animal welfare. Veterinary Record 146, 269-72.

Paul, E.S. and Serpell, J.A. 1996. Obtaining a new pet dog: Effects on middle childhood children and their families. Applied Animal Behaviour Science 47, 17-29.

Paul-Murphy, J., Ludders, J.W., Robertson, S.A., Gaynor, J.S., Hellyer, P.W. and Wong, P.L. 2004. The need for a cross species approach to the study of pain in animals. Journal of the American Veterinary Medical Association 224, 692-7.

Pelat, M., Verwaerde, P., Tran, M.A. Montastruc, J.L. and Senard, J.M. 2002. Alpha (2)-adrenoreceptor function in arterial hypertension associated with obesity in dogs fed a high-fat diet. Journal of Hypertension 20, 957-64.

Pennisi, E. 2004. Genome resources to boost canines' role in gene hunts. Science 304, 1093-4.

Petersen, E.A. 1980. Noise and laboratory animals. Laboratory Animal Care 13, 340-50.

Petersen-Jones, S.M. 1998. A review of research to elucidate the causes of the generalised progressive retinal atropies. The Veterinary Journal 155, 5-18.

Petrie, N., Mellor, D.J., Stafford, K.J., Bruce, R.A. and Ward, R.N. 1996. Cortisol response of calves to two methods of taildocking used with or without local anaesthetic. New Zealand Veterinary Journal 44, 4-8.

Petrie, N., Mellor, D.J., Stafford, K.J., Bruce, R.A. and Ward, R.N. 1996. Cortisol response of calves to two methods of disbudding with or without local anaesthetic. New Zealand Veterinary Journal 44, 9-14.

Pettijohn, T.F., David. K.L. and Scott, J.P. 1980. Influence of living space area on agonistic interaction in telomian dogs. Behavioral and Neural Biology 28, 343-9.

Phipps, N.M. 2003. Rehoming animals from animal rescue shelters in New Zealand. MSc Thesis, Unpublished. Massey University, New Zealand.

Pibarot, P., Dupuis, J., Grisneaux, E., Cuvelliez, S., Plante, J., Beaurgard, G., Bonneau, N.H., Bouffard, J. and Blais, D. 1997. Comparison of ketoprofen, oxymorphone hydrochloride and butorphanol in the treatment of postoperative pain in dogs. Journal of the American Veterinary Medical Association 211, 438-44.

Piercy, R.J., Hinchcliff, K.W., DiSilvestro, R., Reinhart, G.A., Baskin, C.R, Hayek, M.G., Burr, J.R. and Swenson, R.A. 2000. Effects of dietary supplements containing antioxidants on attenuation of muscle damage in exercising sled dogs. American Journal of Veterinary Research 61, 1438-45.

Podberscek, A.L. 1997. Illuminating issues of companion animal welfare through research into human-animal interaction. Animal Welfare 6, 365-72.

Podberscek, A.L. and Blackshaw, J.K. 1993. A survey of dog bites in Brisbane, Australia. Australian Veterinary Practitioner 23, 178-83.

Podberscek, A.L. and Serpell, J.A. 1997. Environmental influences on the expression of aggressive behaviour in English Cocker Spaniels. Applied Animal Behaviour 52, 215-27.

Pongracz, P., Miklosi, A., Doka, A. and Csanyi, V. 2003. Successful application of video-projected human images for signalling in dogs. Ethology 109, 809-21.

Pongracz, P., Miklosi, A., Kubinyi, E., Gurobi, K., Topal, J. and Csanyi, V. 2001. Social learning in dogs. The effect of a human demonstrator on the performance of dogs in a detour task. Animal Behaviour 62, 1109-17.

Posage, J.M., Bartlett, P.C. and Thomas, D.K. 1998. Determining factors for successful adoption of dogs from an animal shelter. Journal of the American Veterinary Medical Association 213, 478-82.

Potthoff, A. and Carithers, R.W. 1989. Pain and analgesia in dogs and cats. Compendium on continuing veterinary education for practicing veterinarians 11, 887-97.

Poulter, D. 1996. Greyhound welfare. Veterinary Record 138, 576.

Poulter, D.A.L. 1981. Injuries to racing greyhounds. Veterinary Record 109, 106.

Prasso, S. 1993. Man's best friend becomes just another tasty morsel. The Veterinarian, XX, 13.

Pratelli, A., Martella, V., Elia, G., Tempesta, M., Guarda, F., Capucchio, M.T., Carmichael, L.E. and Buonavoglia, C. 2001. Severe enteric disease in an animal shelter associated with dual infections by canine adenovirus type 1 and canine coronavirus. Journal of Veterinary Medicine Series B – Infectious Diseases and Veterinary Public Health 48, 385-92.

Prato-Previde, E., Custance, D.M., Spiezio, C. and Sabatini, F. 2003. Is the dog-human relationship an attachment bond? An observational study using Ainsworth's strange situation. Behaviour 140, 225-54.

Price, E.O. 1984. Behavioural aspects of animal domestication. Quarterly Review of Biology 59, 1-27.

Price, E.O. 1999. Behavioural developments in animals undergoing domestication. Applied Animal Behavioural Science 65, 245-271.

Priester, W.A. 1976. Canine intervertebral disc disease – occurrence by age, breed and sex among 8, 117 cases. Theriogenology 6, 293-303.

Prole, J.H.B. 1976. A survey of racing injuries in the Greyhound. Journal of Small Animal Practice 17, 207-18.

Prole, J.H.B. 1981. Greyhound conformation. Veterinary Record 108, 218.

Prole, J.H.B. 1996. Control of oestrus in Greyhounds. Veterinary Record 138, 576.

Pulliainen E. 1975. Wolf ecology in Northern Europe. IN The Wild Canids, edited by MW Fox. Behavioral Science Series. Van Nostrand Reinhold Co, New York, USA.

Rajecki, D.W., Rasmussen, J.L. and Conner, T.J. 2000. Relinquish the dog: Movie messages about misbehaviour. Anthrozoos 13, 140-9.

Rammell, C.G. and Fleming, P.A. 1978. Compound 1080: Properties and use of sodium monofluoroacetate in New Zealand. Ministry of Agriculture and Fisheries, Wellington, New Zealand, 112p.

Ratanakorn, P. 2000. Animal welfare and urban animal management programmes in Thailand. IN Second World Small Animal Veterinary Association Animal Welfare Symposium – Animal welfare issues in urban animal management. Pp. 8

Reilly, J.S. 1993. Euthanasia of animals used for scientific purposes. Australian and New Zealand Council for the care of Animals in Research and Teaching, Adelaide, Australia.

Reisner, L.R., Erb, H.N. and Houpt, K.A. 1994. Risk factors for behaviour-related euthanasia among dominant-aggressive dogs: 110 cases (1989-1992). Journal of the American Veterinary Medical Association 205, 855-63.

Remillard, R.L., Paragon, B.M., Crane, S.W., Debraekeleer, J. and Cowell, C.S. 2000. Making pet food at home. IN Small Animal Clinical Nutrition, edited by M.S. Hand, C.D. Thatcher, R.L. Remillard and P. Roudebush. Mark Morris Institute, Topeka, USA. Pp. 163-181.

Reynolds, A.J. 2000. Diarrhea in a team of sled dogs. IN Small Animal Clinical Nutrition, edited by M.S. Hand, C.D. Thatcher, R.L. Remillard and P. Roudebusch. Mark Morris Institute, Topeka, USA. Pp. 288-9.

Reynolds, A.J., Reinhart, G.A., Carey, D.P., Simmerman, D.A., Frank, D.A. and Kallfelz, F.A. 1999. Effects of protein intake during training on biochemical and performance variables in sled dogs. American Journal of Veterinary Research 60, 789-95.

Rhodes, C.J., Atkinson, R.P.D., Anderson, R.M. and Macdonald, D.W. 1998. Rabies in Zimbabwe: reservoir dogs and the implications for disease control. Philosophical Transactions of the Royal Society of London – Biological Sciences 353, 999-1010.

Richards, J.S., Nepomuceno, C., Riles, M. and Suer, Z. 1982. Assessing pain behaviour: the UAB pain behaviour scale. Pain 14: 393-8.

Roberts, T.D.M. 1954. Cortical activity in electrocuted dogs. Veterinary Record 66, 561-7.

Robertson, I.D. 2003. The association of exercise, diet and other factors with owner-perceived obesity in privately owned dogs from metropolitan Perth, WA. Preventitive Veterinary Medicine 58, 75-83.

Robinson, F.R. and Garner, F.M. 1973. Histopathologic survey of 2500 German shepherd military working dogs. American Journal of Veterinary Research 34, 437-42.

Robinson, J.G.A. and Gorrell, C. 1997. The oral status of a pack of foxhounds fed a 'natural diet'. IN proceedings of the World Veterinary Dental Congress, Birmingham, 35-37.

Roll, A. and Unshelm, J. 1997. Aggressive conflicts amongst dogs and factors affecting them. Applied Animal Behaviour Science 52, 229-42.

Rooney, N.J. and Bradshaw, J.W.S. 2002. An experimental study of the effects of play upon the dog-human relationship. Applied Animal Behaviour Science 75, 161-76.

Rooney, N.J., Bradshaw, J.W.S. and Robinson, I.H. 2000. A comparison of dog-dog and dog-human play behaviour. Applied Animal Behaviour Science 66, 235-48.

Rooney, N.J., Bradshaw, J.W.S. and Robinson, I.H. 2001. Do dogs respond to play signals given by humans. Animal Behaviour 61, 715-22.

Rose, R.J. and Bloomberg, M.S. 1989. Responses to sprint exercise in the Greyhound: effects of haematology, serum biochemistry and muscle metabolites. Research in Veterinary Science 47, 212-8.

Roughan, J.V. and Flecknell, P.A. 2001. Behavioural effects of laparotomy and analgesic effects of ketoprofen and carprofen in rats. Pain 90, 65-74.

Rouhi, A.M. 1997. Detecting illegal substances. C & EN September, 24-9.

Rubin, H.D. and Beck, A.M. 1982. Ecological behaviour of free-ranging urban pet dogs. Applied Animal Ethology 8, 161-8.

Ruble, R.P. and Hird, D.W. 1993. Congenital abnormalities in immature dogs in a pet store: 253 cases (1987-1988). Journal of the American Veterinary Medical Association 202, 633-6.

Rusch, T. 1999. Housing of dogs and cats in animal shelters. Deutsche Tierarzliche Wochenschrift 106, 166-9.

Russell, W.M.S. and Burch, R.L. 1959. The Principles of Human Experimental Technique. Methuen and Co., London, UK.

Sackman, J.E. 1991. Pain: its perception and alleviation indogs and cats. Part I. The physiology of pain. Compendium for Continuing Education of Practicing Veterinarians 13, 71-9.

Sacks, J.J., Lockwood, R., Hornreich, J and Sattin, R.W. 1996. Fatal dog attacks, 1989-1994. Pediatrics 97, 891-5.

Sales, G., Hubrecht, R., Peyvndi, A., Milligan, S. and Shield, B. 1997. Noise in dog kennelling: Is barking a welfare problem for dogs. Applied Animal Behaviour Science 52, 321-9.

Sallander, M., Hedhammar, A., Rundgren, M. and Lindberg, J.E. 2001. Demographic data of a population of insured Swedish dogs measured in a questionnaire survey. Acta Veterinaria Scandanavia 42, 71-80.

Salman, M.D., New, J.G.Jr., Scarlett, J.M., Kass, P.H., Ruch-Gallie, R. and Hetts, S. 1998. Human and animal factors related to the relinquishment of dogs and cats in 12 selected animal shelters in the United States. Journal of Applied Animal Welfare Science 1, 207-26.

Salman, M.D., Hutchison, J., Ruch-Gallie, R., Kogan, L., New, J.C.Jr., Kass, P.H. and Scarlett, J.M. 2000. Behavioural reasons for relinquishment of dogs and cats to 12 shelters. Journal of Applied Animal Welfare Science 3, 93-106.

Salmeri, K.R., Bloomberg, M.S., Sherry, L., Scruggs, B.S. and Shille, V. 1991. Gonadectomy in immature dogs: Effects on skeletal, physical and behavioral development. Journal of the American Veterinary Medical Association 198, 1193-1203.

Sandford, J., Ewbank, R., Molony, V., Tavernor, W.D. and Uvarov, O. 1986. Guidelines for the recognition and assessment of pain in animals. The Veterinary Record 118, 334-8.

Savolainen, P., Zhang, Y., Luo, J., Lundeberg, J. and Leitner, T. 2002. Genetic evidence for an East Asian origin of domestic dogs. Science 298, 1610-3.

Scarlett, J.M., Salman, M.D., New, J.G.Jr. and Kass, P.H. 1999. Reasons for relinquishment of companion animals in U.S. animal shelters: Selected health and personal issues. Journal of Applied Animal Welfare Science 2, 41-57.

Schaefer, J.M., Andrews, R.D. and Dinsmore, J.J. 1981. An assessment of coyote and dog predation on sheep in southern Iowa. Journal of Range Management 45, 883-93.

Schilder, M.B.H. and van der Borg, J.A.M. 2004. Training dogs with help of the shock collar: Short and long term behaviour effects. Applied Animal Behaviour Science 85, 319-334.

Schmutz, S.M. and Schmutz, J.K. 1998. Hereditability estimates of behaviours associated with hunting in dogs. Journal of Heredity 89,233-7.

Schneider, H.P., Truex, R.C. and Knowles, J.O. 1964. Comparative observations of the hearts of mongrel and greyhound dogs. Anatomy Record 149, 173-80.

Schoen, A.M. 2001. Kindred Spirits. Random House Australia, Sydney, Australia.

Schoning, P., Erickson, J. and Milliken, G.A. 1995. Body weight, heart weight, and heart to body weight ratio in Greyhounds. American Journal of Veterinary Research 56, 421-2.

Schuster, A. and Lenard, H.G. 1990. Pain in newborns and prematures: current practice and knowledge. Brain Development 12, 459-65.

Schwartz, S. 2003. Separation anxiety syndrome in dogs and cats. Journal of the American Veterinary Medical Association 222, 1526-32.

Scott, D.W., Miller, W.H. and Griffin, C.E. 1995. Muller and Kirk's Small Animal Dermatology. 5th Edition. WB Saunders Co, Philadelphia, USA.

Scott, J.P. and Fuller, J.L. 1965. Genetics and social behaviour of the dog. The University of Chicago Press, Chicago, USA.

Scott, M.E. 1988. The impact of infection and disease on animal populations: implications for conservation biology. Conservation Biology 2, 40-55.

Seksel, K. and Lindemann, M.J. 2001. Use of clomipramine in treatment of obsessive-compulsive disorder, separation anxiety and noise phobia in dogs: a preliminary clinical study. Australian Veterinary Journal 79, 252-56.

Seksel, K., Mazurski, E.J. and Taylor, A. 1999. Puppy socialisation programs: short and long term behavioural effects. Applied Animal Behavioural Science 62, 335-49.

Selby, L.A., Rhoades, J.D., Irvin. J.A., Carey, G.E. and Wade, R.G. 1980. Values and limitations of pet ownership. Journal of the American Veterinary Medical Association 176, 1274-6.

Serpell, J. 1995. From paragon to pariah: some reflections on human attitudes to dogs. IN The Domestic Dog; its evolution, behaviour and interactions with peoplen edited by J. Serpell. Cambridge University Press, Cambridge, England. Pp. 245-56.

Serpell, J.A. 1996. Evidence for an association between pet behaviour and owner attachment levels. Applied Animal Behaviour Science 47, 49-60.

Serpell, J.A. 2002. Anthropomorphism and anthropomorphic selection – Beyond the "cute" response. Society and Animal 10, 437-54.

Serpell, J. and Jagoe, J.A. 1995. Early experience and the development of behaviour. In The Domestic Dog, its evolution, behaviour and interaction with people, edited by J. Serpell. Cambridge University Press, Cambridge, England. Pp. 79-102.

Sheldon, J.W. 1992. Wild Dogs. Academic Press Inc, San Diego, California, USA.

Sheppard, G. and Mills, D.S. 2003. Evaluation of dog-appeasing pheromone as a potential treatment for dogs fearful of fireworks. Veterinary Record 152, 432-6.

Sibley, K.W. 1984. Diagnosis and management of the overweight dog. British Veterinary Journal 140, 124-31.

Sicard, G.K., Short, K. and Manley, P.A. 1999. A survey of injuries at five greyhound racing tracks. Journal of Small Animal Practice 40, 428-32.

Simpson, B.S. 2000. Canine separation anxiety. Compendium for Continuing Education for Practicing Veterinarians 2, 328-37.

Singer, P. (1998). Practical Ethics, Second Edition. Cambridge University Press, Cambridge, UK.

Sivacolundhu, R.K. 1997. Effects of early desexing in dogs and cats – a preliminary study. MVS Thesis, Faculty of Veterinary Science, University of Melbourne, Australia (unpublished).

Skandakumar, S., Stodulski, G. and Hau, J. 1995. Salivary IgA: A possible stress marker in dogs. A

Slabbert, J.M. and Odendaal, J.S.J. 1999. Early prediction of adult police dog efficiency – a longitudinal study. Applied Animal Behaviour Science 64, 269-88.

Slater, M.R., Robinson, L.E., Zoran, D.L., Wallace, K.A. and Scarlett, J.M. 1995. Diet and exercise patterns in pet dogs. Journal of the American Veterinary Medical Association 207, 186-90.

Slatter, D. 1993. Textbook of small animal surgery. 2nd Edition. WB Saunder Co, Philadelphia, USA.

Slingsby, L.S. and Waterman-Pearson, A.E. 2001. Analgesic effects in dogs of carprofen and pethidine together compared with the effects of either drug alone. Veterinary Record 148, 441-4.

Slingsby, L.S., Jones, A. and Waterman-Pearson, A.E. 2001. Use of a new finger mounted mdevice to compare mechanical nociceptor thresholds in cats given pethidine or no medication after castration. Research in Veterinary Science 70, 243-6.

Sloth, C. 1992. Practical management of obesity in dogs and cats. Journal of Small Animal Practice 33, 178-82.

Snow, D.H., Harris, R.C. and Stuttard, E. 1988. Changes in haematology and plasma biochemistry during maximal exercise in greyhounds. Veterinary Record 123, 487-9.

Soproni, K., Miklosi, J., Topal, J. and Csanti, V. 2001. Comprehension of human communicative signs in pet dogs (Canis familiaris). Journal of Comparative Psychology 115, 122-6.

Soproni, K., Miklosi, J., Topal, J. and Csanti, V. 2002. Dogs (Canis familiaris) responsiveness to human pointing gestures. Journal of Comparative Psychology 116, 27-34.

Soulsby, E.J.L. 1971. Helminths, arthropods and protozoa of domesticated animals. Bailliere, Tindall and Cassell, London.

Staaden, R.V. 1998. Exercise and training. IN Canine sports medicine and surgery, edited by M.S. Bloomberg, J.F. Dee, R.A. Taylor and J.R. Gannon. W.B. Saunders Company, Philadelphia, USA. Pp. 357-63.

Stafford, J.J. 1996. Opinions of veterinarians regarding aggression in different breeds of dogs. New Zealand Veterinary Journal 44, 138-41.

Stafford, K.J., Erceg, V., Kyono, M., Lloyd, J. and Phipps, N. 2003. The dog/human dyad – a match made in heaven. Proceedings of the 28th World Congress of the World Small Animal Veterinary Association. Pp. 225-227.

Steininger, E. 1981. Obesity in dogs. Weiner Tierartzliche Monatschrift 68, 122-30.

Steiss, J., Braund, K., Wright, J., Lenz, S., Heidson, J., Brawner, W., Hathcock, J., Purohit, R., Bell, L. and Horne, R. 1999. Coccygeal muscle injury in English pointers (Limber tail). Journal of Veterinary Internal Medicine 13, 540-8.

Stephens, J.R. 1998. Racing and training greyhounds. IN Canine sports medicine and surgery, edited by M.S.Bloomberg, J.F. Dee, R.A. Taylor and J.R. Gannon. W.B. Saunders Company, Philadelphia, USA. Pp. 369-71.

Stern, M. 1996. Psychological elements of attachment to pets and responses to pet loss. Journal of the American Veterinary Medical Association 209, 1707-11.

Stewart, M.F. 1999. Companion Animal Death. Butterworth Heinemann, Oxford, UK.

Stockman, M. 1997. Frozen tail or limber tail in working dogs. Veterinary Record 140, 588.

Stoone, R.W. 2000. More on ear cropping. Journal of the American Veterinary Medical Association 216, 174.

Strand, P.L. 1993. The pet owner and breeder's perspective on overpopulation. Journal of the American Veterinary Medical Association 202, 922-8.

Stubbs, W.P. and Bloomberg, M.S. 1995. Implications for early neutering in the dog and cat. Seminars in Veterinary Medicine and Surgery (Small Animal) 10, 8-12.

Sturla, K. 1993. Role of breeding regulation laws in solving the dog and cat overpopulation problem. Journal of the American Veterinary Medical Association 202, 928-32.

Sullivan, P.S., Evans, H.L. and McDonald, T.P. 1994. Platelet concentration and haemoglobin function in Greyhounds. Journal of the American Veterinary Medical Association 205, 838-41.

Svobodova, V. 2003. Parasitic infections in an animal shelter. Acta Veterinaria Brno 72, 415-20.

Sweeney, P.A. Greyhound welfare. 1996. Veterinary Record, 138, 504.

Swenson, L. 2001. Population studies on genetic diseases in the dog. Doctoral Thesis, Swedish University of Agricultural Sciences. Sweden.

Swenson, L., Audell, M. and Hedhammer, A. 1997. Prevalence, inheritance and selection for hip dysplasia in seven breeds of dog in Sweden and cost/benefit analysis of a screening and control programme. Journal of the American Veterinary Medical Association 210, 207-14.

Swift, B.J. 2000. Early neutering of dogs. Veterinary Record 147, 667.

Sylvester, S.P., Stafford, K.J., Mellor, D.J., Bruce. R.A. and Ward, R.N. 1998a. Acute cortisol responses of calves to four different methods of dehorning by amputation. Australian Veterinary Journal 76, 123-6.

Sylvester, S.P., Mellor, D.J., Stafford, K.J., Bruce, R.A. and Ward, R.N. 1998a. Acute cortisol responses of calves to scoop dehorning with prior use of local anaesthetic. Australian Veterinary Journal 76, 118-22.

Taddio, A., Goldbach, M., Ipp, M., Stevens, B, and Koren, G. 1995. Effect of neonatal circumcision on pain responses during vaccination in boys. Lancet 345, 291-2.

Takeuchi, Y., Ogata, N., Houpt, K.A. and Scarlett, J.M. 2001. Differences in background and outcome of three behaviour problems in dogs. Applied Animal Behaviour Science 70, 297-308.

Taylor, P. 1985. Analgesia in the dog and cat. In Practice 17, 5-13.

Taylor, P. 2003. Pain management in dogs and cats – More causes and locations to contemplate. The Veterinary Journal 165, 186-7.

Technisearch, 1990. Approaches to pet management for local government. Environment and Technology Policy Unit, Royal Melbourne Institute of Technology, Melbourne, Australia.

Tello, L.H. 2000. The urban pet management in Chile: A perspective from South America. IN Second World Small Animal Veterinary Association Animal Welfare Symposium – Animal welfare issues in urban animal management. Pp. 9-10.

Terlouw, W.M and Lawrence, A.B. 1993. Long-term effects of food allowance and housing on development of stereotypes in pigs. Applied Animal Behaviour Science 38, 103-26.

Thatcher, C.D., Hand, M.S. and Remillard, R.L. 2000. Small animal clinical nutrition: An iterative process. IN Small Animal Clinical Nutrition. Eds M.S. Hand, C.D. Thatcher, R.L. Remillard and P. Roudebush. Mark Morris Institute, Topeka, USA. Pp. 1-20.

Theran, P. 1993. Early-age neutering of dogs and cats. Journal of the American Veterinary Medical Association 202, 914-7.

Thompson, W.R., Melzack, R. and Scott, T.H. 1956. "Whirling behaviour" in dogs related to early experience. Science 123, 939.

Thomson, P.C. 1986. The effectiveness of aerial baiting for the control of dingoes in North-western Australia. Australian Wildlife Research 13, 165-76.

Thomson, P.C. 1992a. The behavioural ecology of dingoes in North-western Australia. I. The Fortescue River study area and details of captured dingoes. Wildlife Research 19, 509-18.

Thomson, P.C. 1992b. The behavioural ecology of dingoes in North-western Australia. II. Activity patterns, breeding season and pup rearing. Wildlife Research 19, 519-30.

Thomson, P.C. 1992c. The behavioural ecology of dingoes in North-western Australia. III. Hunting and feeding behaviour, and diet. Wildlife Research 19, 531-41.

Thomson, P.C. 1992d. The behavioural ecology of dingoes in North-western Australia. IV. Social and spatial organisation and movements. Wildlife Research 19, 543-63.

Thrusfield, M.V., Holt, P.E. and Muirhead, R.H. 1998. Acquired urinary incontinence in bitches: its incidence and relationship to neutering practices. Journal of small animal practice 39, 559-66.

Toll, P.W. 2000a. Poor performance in racing greyhounds. IN Small Animal Clinical Nutrition, edited by M.S. Hand, C.D. Thatcher, R.L. Remillard and P. Roudebusch. Mark Morris Institute, Topeka, USA. Pp. 285-6.

Toll, P.W. 2000b. Weight loss in a cattle dog. IN Small Animal Clinical Nutrition, edited by M.S. Hand, C.D. Thatcher, R.L. Remillard and P. Roudebush. Mark Morris Institute, Topeka, USA. Pp. 286-7.

Toll, P.W. 2000c. Poor performance in a hunting dog. IN Small Animal Clinical Nutrition, edited by M.S. Hand, C.D. Thatcher, R.L. Remillard and P. Roudebusch. Mark Morris Institute, Topeka, USA. Pp. 287-8.

Toll, P.W. and Reynolds, A.J. 2000. The canine athlete. IN Small Animal Clinical Nutrition, edited by M.S. Hand, C.D. Thatcher, R.L. Remillard and P. Roudebusch. Mark Morris Institute, Topeka, USA. Pp. 261-89.

Toll, P.W., Gaehtgens, P., Neuhaus, D., Pieschl, R.L. and Fedde, M.R. 1995. Fluid, electrolyte, and packed cell volume shifts in racing Greyhounds. American Journal of Veterinary Research 56, 227-32.

Topal, J., Miklosi, A., Csanyi, V. and Doka, A. 1998. Attachment behaviour in dogs. Journal of Comparative Psychology 112, 219-29.

Trigg, T.E., Wright, P.J., Armour, A.F., Williamson, P.E., Junaidi, A. and Martin, G.R. 2001. Use of GnRH analogue to produce reversible long-term suppression of reproductive function in male and female domestic dogs. IN Advances in reproduction in dogs, cats and exotic carnivores., edited by P.W. Concannon, G.C.W. England, W. Farstad, C. Linde-Forsberg, J.P. Verstegen and C. Doberska. Journal of Reproductoin and Fertility (Supplement) 57), Portland Press, Essex, England. Pp. 255-61.

Trussel, B.A., King, J. and Smith, D. 1999. Application of environmental enrichment routines to regulatory toxicological studies in the Beagle dog. Animal Technology 50, 131-3.

Trut, L.N. 1999. Early canid domestication: The farm-fox experiment. American Scientist 87, 160-9.

Tuber, D.S., Hennessy, M.B., Sanders, S. and Miller, J.A. 1995. Behavioral and glucocorticoid response of adult domestic dogs (Canis familiaris) to companionship and social separation. Journal of Comparative Psychology 110, 103-8.

Tuber, D.S., Miller, D.D., Caris, K.A., Halter, R., Linden, F. and Hennessy, M.B. 1999. Dogs in animal shelters, suggestions and needed expertise. Psychological Science 10, 379-86.

Tykot, R.H., van der Merwe, N.J. and Hammond, N. 1996. Stable isotope analysis of bone collagen, bone apatite and tooth enamel in the reconstruction of human diet – A study from Cuello, Belize. Archaeological Chemistry ACS Symposium Series 625, 355-65.

Ubbink, G.J., Knol, B.W. and Bouw, J. 1992. The relationship between homozygosity and its occurrence in specific diseases in Bouvier Belge de Flandres dogs in the Netherlands.Veterinary Quarterly 14, 137-40.

Ubbink, G.J., van de Broek, J., Hazelwinkel, H.A. and Rothuizen, J. 1998a. Cluster analysis of the genetic heterogeneity and disease distributions in purebred dog populations. Veterinary Record 142, 209-13.

Ubbink, G.J., van de Broek, J., Hazelwinkel, H.A. and Rothuizen, J. 1998b. Risk estimates for dichotomous genetic disease traits based on a cohort study of relatedness in purebred dogs. Veterinary Record 142, 328-31.

Ubbink, G.J., van den Ingh, T.S., Yuzbasiyan-Gurkan, V., Teske, E., van de Broek, J. and Rothuizen, J. 2000. Population dynamics of inherited copper toxicosis in Dutch Bedlington terriers (1977-1997). Journal of Veterinary Internal Medicine 14, 172-6.

Vaisanen, M., Oksanen, H. and Vainio, O. 2004. Postoperative signs in 96 dogs undergoing soft tissuesurgery. Veterinary Record 155, 729-33.

van der Borg, J.A.M., Netto, W.J.and Planta, D.J.U. 1991. Behaviour testing of dogs in animals shelters to predict behaviour problems. Applied Animal Behaviour Science 32, 237-51.

van Heerden, J. 1989. Small animal problems in developing countries. In Textbook of Veterinary Internal Medicine, edited by SJ Ettiger, 3rd edition WB Saunders Philadelphia Pp. 217-226.

van Oost, B.A. 1998. The role of molecular genetics in the diagnosis of diseases in companion animals: an introduction. Veterinry Quarterly 20 (sullplement) S88-9.

van Valkenburgh, B. and Koepfli, K.P. 1993. Cranial and dental adaptations to predation in canids. Proceedings of a Symposium held by The Zoological Society of London and The Mammal Society. Claredon Press, Oxford, England. Pp. 15-37.

Van Foreest, A.W and Roeters, F.J.M. 1997. Long-term success rate of resin –bonded metal crowns on the canine teeth of working dogs. Veterinary Quarterly 19, 23-28.

Vanhoof, D.R., Deveries, H. and Wensing, J. 1993. Male and female mating competition in wolves – female suppression versus male intervention. Behaviour 127, 141-74.

Veenis, M. 2004. Sterilizing female dogs. Canadian Veterinary Journal 45, 347-8.

Villa, E., Rabano, A., Albarran, O.G., Ruilope, L.M. and Garcia-Robles, R. 1998. Effects of chronic combined treatment with captopril and pravastatin on the progressin of insulin resistance and cardiovascular alterations in an experimental model of obesity in dogs. American Journal of Hypertension 11, 844-51.

Vincent, I.C. and Leahy, R.A. 1997. Real-time non-invasive measurements of heart rate in working dogs: A technique with potential applications in the objective assessment of welfare problems. The Veterinary Journal 153, 179-84.

Vincent, I.C., Mitchell, A.R. and Leahy, R.A. 1993. Non invasive measurements of arterial blood pressure in dogs – a potential indicator for the identification of stress. Research in Veterinary Science 54, 195-201.

Voith, V.L. 1993. Anthropomorphism, spoiling and environmental factors. Tijdschrift voor Diergeneeskunde 118, 61S-3S.

Voith, V.L. 1994. Feeding Behaviour. IN: The Waltham Book of Clinical Nutrition of the Dog and Cat, edited by J.M. Wills and K.W. Simpson. Pergamon, Elsevier Science Ltd, Oxford, UK. Pp. 119-29.

Voith, V.L. and Borchelt, P.L. 1996. Separation anxiety in dogs. IN Readings in Companion Animal Behaviour, edited by V.L. Voith and P.L. Borchelt. Veterinary Learning Systems, New York, USA. Pp. 124-39.

Voith, L.V., Wright, J.C. and Danneman, P.J. 1992. Is there a relationship between canine behaviour and spoiling activities, anthropomorphism and obedience training. Applied Animal Behaviour Science 34, 263-72.

Walker, A. 1997. Orthopaedic injuries in the working dog I. Focus on Farm Working Dogs. Proceedings of the 27th Seminar of the Society of Sheep and Beef Cattle Veterinarians in association with the Companion Animal Society. Foundation for Continuing Education for the New Zealand Veterinary Association, Palmerston North, New Zealand. Pp. 1-10.

Walker, A. 1997. Orthopaedic injuries in the working dog II. Focus on Farm Working Dogs. Proceedings of the 27th Seminar of the Society of Sheep and Beef Cattle Veterinarians in association with the Companion Animal Society. Foundation for Continuing Education for the New Zealand Veterinary Association, Palmerston North, New Zealand. Pp. 11-25.

Wall, P.D. 1992. Defining pain in animals. IN Animal Pain, edited by C.E. Short and A. van Podnak. Churchill Livingston, New York, USA. Pp. 63-79.

Wansbrough, R.K. 1996. Cosmetic tail docking of dogs. Australian Veterinary Record 74, 59-63

Ward, M.R. 2003. Behavioural therapy success and the effect of socialisation on subsequent behaviour in dogs. MVS Thesis. Unpublished. Massey University, New Zealand.

Ward, M.P., Glickman, L.T. and Guptill, L.F. 2002. Prevalence and risk factors for leptospirosis among dogs in the United States and Canada: 677 cases (1970-1988). Journal of the American Veterinary Medical Association 220, 53-58.

Warman, B. 2004. Lets talk tails. New Zealand Kennel Gazette XXX, 12-13.

Watson, A.D.J., Nicholson, A., Church, D.B. and Pearson, M.R.B. 1996. Use of anti-inflammatory agents and analgesic drugs in dogs and cats. Australian Veterinary Journal 74, 203-10.

Watson, D. 1996. Longevity and diet. Veterinary Record 138, 71.

Wayne, R.K. 1993. Molecular evolution of the dog family. Trends in Genetics 9, 218-24.

Wayne, R.K. 2001. Consequences of domestication: Morphological diversity in the dog. IN The Genetics of the Dog, edited by A. Ruvinsky and J. Sampson. CABI Publishing, England.

Webb, A.A. 2003. Potential sources of neck and back pain in clinical conditions of dogs and cats: A review. The Veterinary Journal 165, 193-213.

Webster, J.M. 1992. Docking of dogs. Veterinary Record 131, 374.

Wells, D. and Hepper, P.G. 1992. The behaviour of dogs in a rescue shelter. Animal Welfare 1, 171-86.

Wells, D.L. 2001. The effectiveness of a citronella spray collar in reducing certain forms of barking in dogs. Applied Animal Behaviour Science 73, 299-309.

Wells, D.L. 2003. Comparison of two treatments for preventing dogs eating their own faeces. Veterinary Record 153, 51-3.

Wells, D.L. 2004. A review of environmental enrichment for kennelled dogs, *Canis familiaris*. Applied Animal Behaviour Science 85, 307-17.

Wells, D.L. and Hepper, P.G. 1998. A note on the influence of visual conspecific contact on the behaviour of sheltered dogs. Applied Animal Behaviour Science 60, 83-8.

Wells, D.L. and Hepper, P.G. 1999. Male and females dogs respond differently to men and women. Applied Animal Behaviour Science 61, 341-9.

Wells, D.L. and Hepper, P.G. 2000a. The influences of environmental change on the behaviour of sheltered dogs. Applied Animal Behaviour Science 68, 151-62.

Wells, D.L. and Hepper, P.G. 2000b. Prevalence of behaviour problems reported by owners of dogs purchased from an animal rescue shelter. Applied Animal Behaviour Science 69, 55-65.

Wells, D.L., Graham, L. and Hepper, P.G. 2002. The influence of auditory stimulation on the behaviour of dogs housed in a rescue shelter. Animal Welfare 11, 385-93.

Welsh, E..M. and Nolan, A.M. 1995. The effect of abdominal surgery on thresholds to thermal and mechanical stimulation in sheep. Pain 60, 159-66.

World Health Organisation. 1994. Oral vaccines against canine rabies: WHO announces world's first field trials.

Wilkins, C.M. 1997. Frozen tail or limber tail in working dogs. Veterinary Record 140, 588.

Wilkins, D., Mews, A. and Bate, T. 1988. Illegal dog fighting. Veterinary Record 122, 310.

Wilkinson, L. 1992. Animals and disease. Cambridge University Press, UK.

Wilkinson, M.J. and McEwan, N.A. 1991. Use of ultrasound in the measurement of subcutaneous fat and prediction of total body fat in dogs. Journal of Nutrition 121, S47-S50.

William, V.M., Lascelles, B.D.X. and Robson, M.C. 2005. Current attitudes to, and use of, peri-operative analgesia indogs and cats by veterinarians in New Zealand. New Zealand Veterinary Journal 53, 193-202.

Willis, M.B. 1995. Genetic aspects of dog behaviour with particular reference to working ability. IN The Domestic Dog; its evolution, behaviour and interaction with people, ediyed by J. Serpell. Cambridge University Press, Cambridge, England. Pp. 51-64.

Willis, M.B. 1997. A review of the progress in canine hip dysplasia control in Britain. Journal of the American Veterinary Medical Association 210, 1480-2.

Wilson, P.R. and Stafford, K.J. 2002. Welfare of farmed deer in New Zealand. 2 Velvet Antler Removal. New Zealand Veterinary Journal 50, 221-7.

Wilsson, E. and Sundgren, P. 1997. The use of a behaviour test for selection of dogs for service and breeding. II. Heritability for tested parameters and effects of selection based on service dog characteristics. Applied Animal Behaviour Science 54, 235-41.

Wilsson, E. and Sundgren, P.E. 1998a. Behaviour tests for 8 week old puppies – heritabilities of tested behaviour traits and its correspondence to later behaviour. Applied Animal Behaviour Science 58, 151-62.

Wilsson, E. and Sundgren, P.E. 1998b. Effects of weight, litter size and parity of mother on the behaviour of the puppy and the adult dog. Applied Animal Behaviour Science 58, 245-54.

Wisely, S.M., Ososky, J.J. and Buskirk, S.W. 2002. Morphological changes to black-footed ferrets (*Mustela nigripes*) resulting from captivity. Canadian Journal of Zoology 80, 1562-8.

Wismer, T.A., Murphy, L.A., Gwaltney-Brant, S.M. and Albretsen, J.C. 2003. Management and prevention of toxicosis in search-and–rescue dogs responding to urban disasters. Journal of the American Veterinary Medical Association 222, 305-9.

Wolfle, T.L. 1987. Control of stress using non drug approaches. Journal of the American Veterinary Medical Association 191, 1219-21.

Wolfle, T.L. 1990. Policy, program and people: The three Ps to well being. IN Canine Research Environment, edited by J.A. Murphy and L. Krulisch. Scientists Centre for Animal Welfare, Bethswden, MD, USA. Pp. 41-47.

Wood, J.L.N. and Lakhani, K.H. 1998. Deafness in Dalmatians: Does sex matter? Preventive Veterinary Medicine 36, 39-50.

Worth, A.J., Ainsworth, S.J., Brocklehurst, P.J. and Collett, M.G. 1997. Nitrite poisoning in cats and dogs. New Zealand Veterinary Journal 45, 193-5.

Yeon, S.C., Golden, G., Sung, W., Erb, H.N., Reynolds, A.J.and Houpt, K.A. 2001. A comparison of tethering and pen confinement of dogs. Journal of Applied Animal Welfare 4, 257-70.

Zawistowski, S., Morris, J., Salman, M.D. and Ruch-Gallie, R. 1998. Population dynamics, overpopulation, and the welfare of companion animals: New insights into old and new data. Journal of Applied Animal Welfare 3, 193-206.

Whay, H.R., Waterman, A.E. and Webster, A.J.F. (1997). Associations between locomotion, claw.

Whay, H.R., Waterman, A.E., Webster, A.J.F. and O'Brien, J.K. (1998). The influence of lesion type on the duration of hyperalgesia associated with hindlimb lameness in dairy cattle.VeterinaryJournal 156, 23-9.

WHO. 1994. Oral vaccines against canine rabies: WHO announces world's first field trials.

Wilkins, C.M. 1997. Frozen tail or limber tail in working dogs. Veterinary Record 140, 588.

Wilkins, D., Mews, A. and Bate, T. 1988. Illegal dog fighting. Veterinary Record 122, 310.

Willis, M.B. 1995. Genetic aspects of dog behaviour with particular reference to working ability. IN The Domestic Dog; its evolution, behaviour and interaction with people. Ed J. Serpell. Cambridge University Press, Cambridge, England. Pp. 51-64.

Willis, M.B. 1997. A review of the progress in canine hip dysplasia control in Britain. Journal of the American Veterinary Medical Association 210, 1480-2.

Wilsson, E. and Sundgren, P. 1997. The use of a behaviour test for selection of dogs for service and breeding. II. Heritability for tested parameters and effects of selection based on service dog characteristics. Applied Animal Behaviour Science 54, 235-41.

Wilsson, E. and Sundgren, P.E. 1998a. Behaviour tests for 8 week old puppies – heritabilities of tested behaviour traits and its correspondence to later behaviour. Applied Animal Behaviour Science 58, 151-62.

Wilsson, E. and Sundgren, P.E. 1998b. Effects of weight, litter size and parity of mother on the behaviour of the puppy and the adult dog. Applied Animal Behaviour Science 58, 245-54.

Wisely, S.M., Ososky, J.J. and Buskirk, S.W. 2002. Morphological changes to black-footed ferrets (*Mustela nigripes*) resulting from captivity. Canadian Journal of Zoology 80, 1562-8.

Wismer, T.A., Murphy, L.A., Gwaltney-Brant, S.M. and Albretsen, J.C. 2003. Management and prevention of toxicosis in search-and–rescue dogs responding to urban disasters. Journal of the American Veterinary Medical Association 222, 305-

Wolfsheimer, K.J. 1994. Obesity in dogs. Compendium for the continuing education of practicing veterinarians 16, 981-XX.

Wood, J.L.N. and Lakhani, K.H. 1998. Deafness in Dalmatians: Does sex matter? Preventive Veterinary Medicine 36, 39-50.

Young, L.A., Dodge, J.C., Guest, K.J., Cline, J.L. and Kerr, W.W. 2002. Age, breed, sex and period effects on skin biophysical parameters for dogs fed canned dog food. Journal of Nutrition 132, 1695S-7S.

Yuon, S.C., Golden, G., Sung, W., Erb, H.N., Reynolds, A.J.and Houpt, K.A. 2001. A comparison of tethering and pen confinement of dogs. Journal of Applied Animal Welfare 4, 257-70.

INDEX

W

Z